Fabrication of Long-Length and Bulk High-Temperature Superconductors

Related titles published by The American Ceramic Society

Ceramic Materials and Multilayer Electronic Devices (Ceramic Transactions, Volume 150)
Edited by K.M. Nair, A.S. Bhalla, S.-I. Hirano, D. Suvorov, W. Zhu, and R. Schwartz
©2004, ISBN 1-57498-205-2

Morphotropic Phase Boundary Perovskites, High Strain Piezoelectrics, and Dielectric Ceramics (Ceramic Transactions Volume 136)
Edited by Ruyan Guo, K. M. Nair, Winnie K. Wong-Ng, Amar Bhalla, Dwight Viehland, D. Suvorov, Carl Wu, and S-I. Hirano
©2003, ISBN 1-57498-151-158

Processing of High-Temperature Superconductors (Ceramic Transactions, Volume 140)
Edited by Amit Goyal, Winnie Wong-Ng, Masato Murakami, and Judith Driscoll
©2003, ISBN 1-57498-155-2

Recent Developments in Electronic Materials and Devices (Ceramic Transactions, Volume 131)
Edited by K.M. Nair, A.S. Bhalla, and S.-I. Hirano,
©2002, ISBN 1-57498-145-5

Dielectric Materials and Devices
Edited by K.M. Nair, Amar S. Bhalla, Tapan K. Gupta, Shin-Ichi Hirano, Basavaraj V. Hiremath, Jau-Ho Jean, and Robert Pohanka
©2002, ISBN 1-57498-118-8

The Magic of Ceramics
By David W. Richerson
©2000, ISBN 1-57498-050-5

Electronic Ceramic Materials and Devices (Ceramic Transactions, Volume 106)
Edited by K.M. Nair and A.S. Bhalla
©2000, ISBN 1-57498-098-X

For information on ordering titles published by The American Ceramic Society, or to request a publications catalog, please contact our Customer Service Department at:

Customer Service Department
PO Box 6136
Westerville, OH 43086-6136, USA
614-794-5890 (phone)
614-794-5892 (fax)
info@ceramics.org

Visit our on-line book catalog at www.ceramics.org.

Ceramic Transactions
Volume 149

Fabrication of Long-Length and Bulk High-Temperature Superconductors

Proceedings of the Fabrication of Long-Length and Bulk High-Temperature Superconductors held at the 105th Annual Meeting of The American Ceramic Society, April 27-30, 2003, in Nashville, Tennessee

Edited by

Ruling Meng
University of Houston

Amit Goyal
Oak Ridge National Laboratory

Winnie Wong-Ng
National Institute of Standards and Technology

Kaname Matsumoto
Kyoto University

Herbert C. Freyhardt
Institut fur Matellphysik, Universitat Gottingen

Published by
The American Ceramic Society
PO Box 6136
Westerville, Ohio 43086-6136
www.ceramics.org

Proceedings of the Fabrication of Long-Length and Bulk High-Temperature Superconductors held at the 105th Annual Meeting of The American Ceramic Society, April 27-30, 2003, in Nashville, Tennessee

COVER PHOTO: "Cross sectional view of the 19 filament hexagonal tube with $SrZrO_3$ oxide barrier" is courtesy of Se-Jong Lee, Deuk Yong Lee, Yo-Seung Song, and Kyung-Hwan Ye and appears as figure 2 in their paper "Preparation of $SrZrO_3$ Thin Films on Bi(2223) Tapes for the Reduction in AC Losses," which begins on page 95.

For information on ordering titles published by The American Ceramic Society, or to request a publications catalog, please call 614-794-5890.

4 3 2 1–07 06 05 04
ISSN 1042-1122
ISBN 1-57498-204-4

Contents

BSCCO–Based Conductors, Mgb$_2$ and Other HTS Materials

Control of Microstructure

Preface

The discovery of high-temperature superconductivity (HTS) 16 years ago has been hailed as one of the most exciting advancements in modern physics. However, the full impact of HTS will only be realized with the success of large-scale applications. For most large-scale bulk applications of HTS, long-length flexible wires that can carry a large amount of supercurrent are required. During the last decade, remarkable progress has been made in processing HTS conductors for large-scale applications. For example, kilometer-length Bi-containing (BSCCO) first-generation wires/tapes have been successfully fabricated. Significant improvement in the performance of the $YBa_2Cu_3O_x$ (YBCO) - based second-generation wires/tapes (coated conductors) has also been achieved in short-length (up to 10 meters) tapes. Substrate fabrication techniques that have received considerable attention include IBAD (ion-beam-assisted deposition), RABiTS (rolling assisted bi-axially textured substrates), and ISD (inclined substrate deposition). Epitaxial growth of YBCO and other oxides on these substrates can be accomplished by a variety of techniques including pulsed laser deposition (PLD), electron beam evaporation, sputtering, chemical combustion vapor deposition (CCVD), jet vapor deposition, ex-situ BaF_2 processing, ex-situ sol-gel techniques, and liquid phase epitaxy (LPE). A worldwide research target is the development of practically scalable techniques for the fabrication of long-length, YBCO thick films. Another major focus area is that of the growth of thick YBCO films without massive degradation of the superconducting properties.

For bulk applications, large crystal-like superconductor pucks, which are often produced by melt-textured techniques, are required. Applications of melt-textured YBCO materials, which exhibit large-domain levitation, include frictionless bearings for flywheels, contact-less transportation, damping devices, flux-trap magnets, magnetic shields, and current leads. Various important processing issues concerning melt-textured HTS are still not yet fully understood. Furthermore, strategic research on microstructural issues such as weak-links, flux pinning, grain boundaries, and phase diagrams is still needed for the development of materials in both bulk and wire/tape form.

This proceedings volume contains papers presented at the Fabrication of Long-Length & Bulk HTS Conductors symposium during the 105th Annual Meeting of The American Ceramic Society (ACerS), April 27-30, 2003 in Nashville, Tennessee. The symposium featured the above-mentioned issues pertaining to the processing of HTS materials, and also the current status and potential of the YBCO-based coated conductors and the BSCCO-based conductors (especially Bi-2223 and Bi-2212) already in production. MgB_2 based wires for lower temperature applications were also highlighted. The papers are divided into three sections: (1) coated conductors, (2) BSCCO-based conductors, MgB_2 , and other HTS materials, and (3) control of microstructure. The order in which the papers appear in this proceedings volume and the division into which they are organized may be different from that of their presentation at the meeting. It is hoped that this comprehensive volume will be a good summary of the latest developments in high-temperature superconductor research as well as good source material for researchers and managers working in this field.

We acknowledge the service provided by the session chairs and appreciate the valuable assistance from the ACerS programming coordinators. We are also in debt to Greg Geiger and Bill Jones for their involvement in editing and producing this book. Special thanks are due to the speakers, authors, manuscript reviewers, and ACerS officials for their contributions.

Ruling Meng

Amit Goyal

Winnie Wong-Ng

Kaname Matsumoto

H. C. Freyhardt

Coated Conductors

SOLUTION BUFFER LAYERS FOR YBCO COATED-CONDUCTORS

S. Sathyamurthy, M. Paranthaman, H-Y.Zhai, S.Kang, C.Cantoni, S.Cook, L. Heatherly, A. Goyal, and H.M.Christen
Oak Ridge National Laboratory, Oak Ridge, TN 37831
Md.S. Bhuiyan, and K. Salama
University of Houston, Houston, TX 77204.

A single layer of $La_2Zr_2O_7$ (LZO), deposited on textured Ni and Ni-3%W (Ni-W) tapes by a low-cost sol-gel process, is used as a buffer layer for the growth of $YBa_2Cu_3O_{7-\delta}$ (YBCO) coated conductors. It is shown for the first time that such single buffer layers can be used for the deposition of YBCO yielding critical current densities (J_c) that are comparable to those typically obtained using $CeO_2/YSZ/Y_2O_3$ tri-layers on identical substrates, i.e. in excess of 1 MA/cm^2 on nickel substrates and close to 2 MA/cm^2 on Ni-W substrates (which has a sharper texture) at 77K and self-field. Reel-to-reel deposition of LZO buffer layers in long lengths and their use in fabrication of coated conductors in long lengths has also been demonstrated. These results offer promise to the fabrication of low cost all-solution coated conductors.

INTRODUCTION

The focus of our research in the area of high-temperature superconductivity (HTS) in recent years has been on the development of the second generation wires also known as coated-conductors [1-3]. One of the leading textured-template approaches for the fabrication of coated-conductors is the rolling assisted biaxially textured substrates (RABiTS) approach [2]. In this approach, cube textured nickel or Ni alloy substrates, obtained by cold rolling and recrystallization, act as a template for the epitaxial deposition of buffer layers and the YBCO superconductor. The buffer layers, apart from providing a structural template, also acts as a chemical barrier between the metal substrate and the HTS coating. Using such architecture, sufficient biaxial texturing of the HTS layer has been obtained to avoid problems associated with weak-linked, high-angle grain boundaries [4].

To date, in the processing of high current coated conductors using RABiTS, the best results have been obtained reproducibly using three or four layer buffer architectures like $CeO_2/YSZ/Y_2O_3/Ni/NiW$. The fabrication of this multi-layered buffer architecture, however, may present significant roadblocks to the scale-up of the process to long-lengths. Typically, deposition of the buffer layer stack would involve a combination of e⁻ beam and rf-sputter deposition techniques coupled with the exposure of the samples to thermal cycling and ambient environment. These requirements could lead to the degradation of the individual buffer layers and add to the complexity, and cost of the over-all process. However, if a single buffer layer deposited using a scaleable technique is developed, it would significantly decrease the processing time and make the process simpler and more conducive to scale-up to long lengths.

EXPERIMENTAL

Stoichiometric quantities of lanthanum isopropoxide and zirconium-n-propoxide were dissolved in 2-methoxyethanol and refluxed to obtain the precursor solution with 0.25 M cation concentration. All-solution buffered substrates were prepared by spin coating the precursor solution on textured Ni and Ni-3at. % W (NiW) substrates, and heat treating the films at 1100°C for 1 h in Ar/4%H$_2$ atmosphere (described in detail elsewhere [5]). The process was repeated to get thicker coating of the buffer layer. For long length samples, the buffer layer deposition was carried out in a reel-to-reel dip coating process, and the samples were also heat treated in a continuous reel-to-reel fashion. The YBa$_2$Cu$_3$O$_x$ (YBCO) was deposited using pulsed laser deposition (PLD) at 790°C in 120mTorr oxygen with average laser energy of 400-410mJ. The phase purity and texture of the samples was analyzed using X-ray diffraction (XRD). The film thickness and compositional homogeneity were analyzed using Rutherford Backscattering Spectroscopy (RBS) and the microstructural analysis of the samples was performed using a Scanning Electron Microscope (SEM). The critical current densities of the samples were measured using a standard four-point probe technique with a 1 μV/cm voltage criterion.

RESULTS AND DISCUSSION

A typical XRD of multiple coats of LZO on NiW substrates is shown in Fig.1. This figure illustrates the presence of a strong (400) reflection from the LZO film while the (222) reflection is at the background levels. It is clear from this figure the highly oriented LZO films can be processed using multiple coating and annealing cycles. On the multiple coated LZO films, YBCO was deposited by PLD. Figure 2 shows the XRD pattern of the YBCO film on LZO buffered NiW substrates. Along with a good c-axis texture, the XRD pattern also shows that there is no detectable amount of NiO in the sample. This suggests that the LZO buffer layer acts as a good barrier layer and protects the NiW substrate from oxidation. The microstructure of the multiple coated LZO films and the PLD deposited YBCO films are shown in Fig. 3. This figure shows that the LZO films are fine-grained, dense and crack-free with a smooth surface. Such a dense, smooth surface is essential for all-solution buffers to be effective. The YBCO microstructure is typical of what is observed for PLD films on metal substrates.

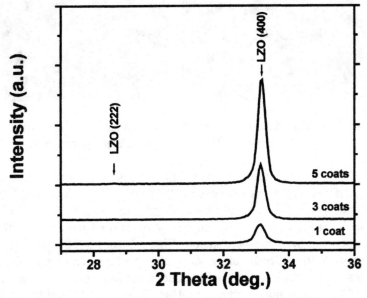

Fig.1 XRD patterns of multiple coats of LZO on textured Ni substrates.

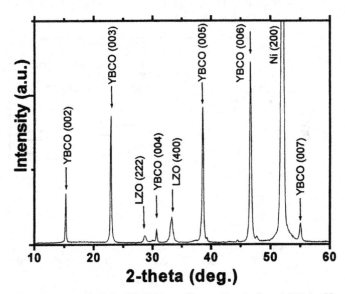

Fig.2 XRD pattern of PLD-YBCO on 60nm all-solution LZO buffered Ni substrates.

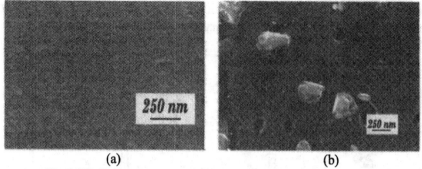

(a)   (b)

Fig.3 SEM Microstructures of (a) 3 coats LZO on Ni substrate,
(b) PLD-YBCO on LZO buffered Ni substrate.

Using RBS, the films were analyzed for thickness, composition and the quality of the interfaces. From this analysis, the thickness of the YBCO and LZO films were found to be 200nm and 60nm respectively, and both YBCO and LZO were found to be stoichiometric. This analysis also showed that the interfaces and surfaces between the films were smooth. Figure 4 compares the field dependence of the critical current density of the all-solution buffered samples on Ni and NiW substrates is compared with the corresponding performance of all-vacuum three layer buffers on similar substrates. This figure clearly illustrates that the performance of all-solution LZO buffers is comparable to that of the three layer buffers. Critical current density of over 1 MA/cm^2 on Ni substrates and about 2 MA/cm^2 on NiW substrates has been measured.

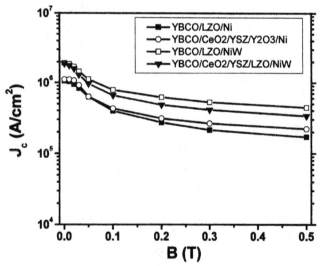

Fig. 4 Comparison of the J_c Vs B performance of YBCO on all-solution LZO buffers that of YBCO on tri-layer buffer architectures for Ni and NiW substrates.

Using a reel-to-reel dip coating process (described elsewhere [5]), multiple coats of LZO were processed over long lengths of Ni and NiW substrates. The in-plane and out-of-plane texture of 3 coats of LZO on Ni substrate and 4 coats of LZO on NiW substrate were characterized using a reel-to-reel X-ray diffractometer. The results from these measurements are shown in Fig.5. It is clear from this figure that using a reel-to-reel dip coating process, it is possible to deposit multiple coated LZO films with uniform texture over the entire length of the tape. It also illustrates that the texture of the LZO film is as good as that of the metal substrate. This offers promise to processing long lengths of solution buffered Ni and NiW tapes for coated conductor fabrication.

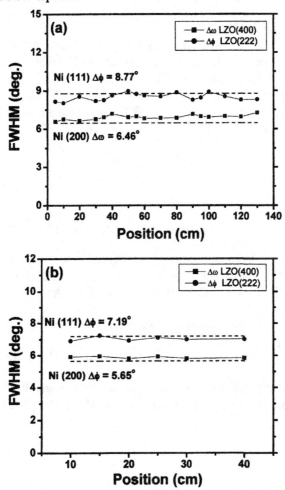

Fig. 5 Texture data from multiple coated LZO films on long length metal substrates. (a) 3-coats of LZO on long Ni substrate and (b) 4-coats of LZO on long NiW substrate.

CONCLUSIONS

A solution-based approach to buffer layer processing has been explored for coated conductors fabrication. Critical current densities over 1 MA/cm^2 on Ni substrates and up to 2 MA/cm^2 on NiW substrates have been measured using all-solution LZO buffer layers. Using XRD, it has been confirmed that the LZO films can act as good barrier layers and prevent the metal substrates from oxidation. Multiple coating of LZO has also been demonstrated on long lengths of Ni and NiW substrates with uniform texture through the length of the tape. These results are promising and give hope to the possibility of developing all-solution coated-conductors in long lengths.

ACKNOWLEDGEMENTS

The U.S. DOE, Division of Materials Sciences, Office of Science and Office of Power Technologies-Superconductivity Program, and Office of Energy Efficiency and Renewable Energy sponsored this research. This research was performed at the Oak Ridge National Laboratory, managed by UT-Battelle, LLC for the U.S.DOE under contract DE-AC05-00OR22725. Sincere acknowledgements are also extended to Oak Ridge Associated Universities for making this work possible.

REFERENCES

[1] Y. Iijima, N. Tanabe, O. Kohno and Y. Ikeno, Appl. Phys. Lett., **60**, 769 (1992).

[2] A. Goyal, D. P. Norton, J. D. Budai, M. Paranthaman, E. D. Specht, D. M. Kroeger, D. K. Christen, Q. He, B. Saffian, F. A. List, D. F. Lee, P. M. Martin, C. E. Klabunde, E. Hartfield, and V. Sikka, Appl. Phys. Lett., **69**, 1795 (1996).

[3] M. Bauer, R. Semerad, and H. Kinder, IEEE Trans. Appl. Supercond., **9**, 1502 (1999).

[4] D. Dimos, P. Choudhary, J. Mannhart, and F.K. Le Goues, Phys. Rev. Lett. **61**, 219 (1988).

[5] S. Sathyamurthy, M. Paranthaman, T. Aytug, B.W. Kang, P. M. Martin, A. Goyal, D.M. Kroeger, and D. K. Christen, Jl. Mater. Res., **17**, 1543 (2002).

Fabrication of High Temperature Superconductor

SCALE UP OF HIGH PERFORMANCE HIGH TEMPERATURE SUPERCONDUCTORS

V. Selvamanickam, Y. Li, H. G. Lee, X. Xiong, Y. Qiao, J. Reeves, Y. Xie, A. Knoll, and K. Lenseth
SuperPower Inc., Schenectady, NY 12304

ABSTRACT
    Second-generation High Temperature Superconducting (HTS) conductors have been fabricated by two high rate processes, Pulsed Laser Deposition (PLD), and Metal Organic Chemical Vapor Deposition (MOCVD). Ten meter long tapes have been produced with performance exceeding 100 A end to end. Also, critical currents as high as 173 A have been achieved in meter-long lengths by MOCVD.

INTRODUCTION
    Second-generation HTS conductors based on RE-Ba-Cu-O (RE = rare-earths) have been developed since the early 1990's for electric power applications. The enabling technology for the second-generation HTS conductors is a textured template that can be produced by a variety of techniques. These include Ion Beam Assisted Deposition (IBAD) [1-3], Rolling Assisted Biaxially Textured Substrates (RABiTS) [4], and Inclined Substrate Deposition (ISD) [5]. Several techniques are being used to deposit HTS layers on the textured templates including PLD, MOCVD, E-beam Evaporation (BaF$_2$ process), and Metal Organic Deposition. Among these process PLD and MOCVD offer the advantage of high deposition rates. Deposition rates greater than 100 Angstroms/second have been demonstrated to produce YBCO films with Jc greater than 1 MA/cm^2 by both PLD [6] and MOCVD [7]. In addition to high deposition rate, MOCVD offers an additional advantage of large deposition zone. The deposition zone in MOCVD can be as long and as wide as the showerhead. Because of the combination of high deposition rate and large deposition zone, very high tape throughput can be achieved with MOCVD. We have been scaling up both high rate processes, PLD and MOCVD to fabricate second-generation HTS conductors in long lengths. In this article, we report the progress in high performance second-generation HTS conductors based on PLD and MOCVD.

EXPERIMENTAL AND RESULTS

Hastelloy C-276 substrates are polished in a custom built Chemical Mechanical Polishing facility in lengths of up to 100 m in a single pass. The surface roughness of the polished substrates is monitored on-line in the polishing facility using a laser scatterometer device. Measurements are obtained every 0.1 mm of the polished tape. Measurement obtained over a 10 m long section of a polished tape is shown in figure 1. As shown in figure 1, the average surface roughness over 10 m is about 1 nm and the standard deviation is approximately 0.08 nm. This roughness value compares well with the values we have reported on meter-long substrates [8].

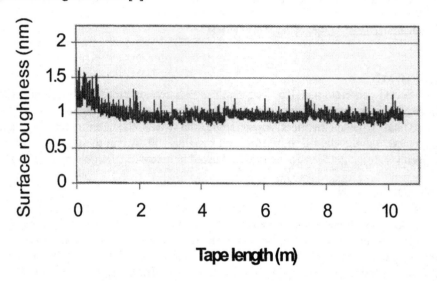

Figure 1 Surface roughness measurements obtained on-line from a 10 m long substrate tape polished by chemical mechanical polishing.

Buffer layers are then deposited on the polished substrate in a pilot IBAD facility. This facility is equipped with r.f. ion sources that can be operated continuously for several hundred hours. Ten meter long buffered substrates are routinely produced in the pilot IBAD facility. The in-plane texture of the buffered substrates is measured in a General Area Detector Diffraction System (GADDS) equipped with a spooling system. Direct polefigures of the film are obtained over long tape lengths. This avoids problems that can arise from indirectly determining the texture quality from intensity measurements. Results from direct in-plane texture measurements over a 10 m long buffered IBAD substrate are shown in figure 2. The texture measurement was obtained every meter of the tape. The average texture over 10 m is about 10.9 degrees with a standard deviation of 0.7 degrees. This compares well with the best reported bulk texture values on IBAD tapes. It also compares well with the texture values that we obtained on meter long tapes [9].

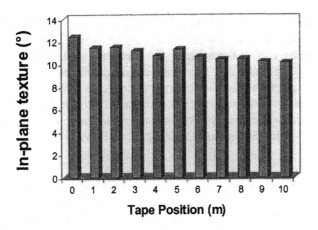

Figure 2 Distribution of in-plane texture in a 10 m long buffer tape produced by IBAD

Deposition of superconducting layers is conducted in a pilot PLD facility and a prototype MOCVD facility. The PLD facility is equipped with three, 3" diameter targets that can be rotated as well as rastered for uniform ablation. An industrial laser, Lambda Steel 670, that is rated for 24 hours/day, 7 days/week operation is used. This laser can deliver a maximum pulse energy of 670 mJ/pulse at a maximum repetition rate of 300 Hz. Pulse repetition rates of up to 200 Hz have been employed in this work. Previously, we reported achievement of critical currents over 100 A and up to 135 A with multiple meter-long tapes [10]. We also reported demonstration of 106 A over a 3 m long tape produced using PLD. In this article, we report demonstration of 106 A over 10 m long conductor. Results from critical current measurements from this tape are shown in figure 3.

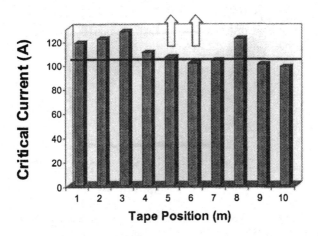

Figure 3 Critical current distribution over a 10 m PLD tape that exhibited an end-to-end critical current value of 106 A.

Critical current of the tape was measured every 1 m using a four-probe technique and a voltage criterion of 1 microvolt/cm. Every meter-long section of the tape exhibited a critical current over 100 A except the last meter (98 A). Also, as indicated by 2 arrows in fig. 3, critical current measurements were halted in 2 sections before complete transition because of concerns with tape heating. To our knowledge, this is only the second demonstration of performance over 100 A in a 10 m long second-generation HTS conductor produced by PLD. Also, this is likely the first such demonstration using a scaleable reel-to-reel deposition system.

We have previously described in detail the MOCVD facility and process that we developed to fabricate high performance second-generation HTS conductors [11]. Our MOCVD process uses a liquid precursor delivery system which has the advantages of long-term stability, high throughput, and reproducibility. We demonstrated 100 A and 1 MA/cm^2 performance in short samples of second-generation conductor using MOCVD a few years ago [11]. Recently, we scaled up our process to demonstrate multiple meter-long tapes with performance exceeding 100 A [10]. We believe that was the first demonstration of meter-long second-generation HTS conductor with performance greater than 100 A using MOCVD. In this work, we further improved the performance of meter-long conductors produced by MOCVD. Figure 4 shows a current-voltage curve obtained from a meter-long tape produced by MOCVD. As shown in the figure, a critical current of 173 A, corresponding to a Jc of 1.2 MA/cm^2 was achieved. To our knowledge there has been only one other demonstration of a meter-long second-generation HTS conductor with a better performance. That demonstration was with PLD [12].

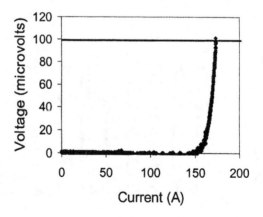

Figure 4 Current-voltage curve obtained from a 1 m long tape produced by MOCVD. A critical current value of 173 A was obtained in this tape.

Fabrication of High Temperature Superconducto

Figure 5 exhibits the critical current distribution over a 2.8 m long tape produced by MOCVD. An end-to-end critical current of 105 A was achieved in this tape. Critical current measurements were obtained every 0.2 m of the tape and every section exhibited a critical current greater than 100 A except one (97 A). As shown by the arrows in figure 6, measurements had to be halted in 8 sections because of tape heating.

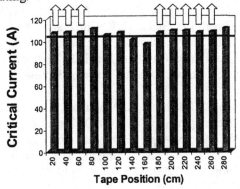

Figure 5 Critical current distribution over a 2.8 m MOCVD tape that exhibited an end-to-end critical current value of 105 A.

SUMMARY
Up to 10 m long second-generation HTS conductors have been produced. Surface roughness values of 1 nm have been achieved uniformly over 10 m long Hastelloy-C substrates by reel-to-reel chemical mechanical polishing. In-plane texture values of 10.9 degrees have been achieved uniformly over 10 m long buffer layers deposited by IBAD. End-to-end critical current value of 106 A has been demonstrated in 10 m long conductor produced by PLD. MOCVD has been used to produce second-generation conductor with critical currents as high as 173 A in meter-lengths and 105 A in 2.8 m lengths.

ACKNOWLEDGMENTS
This work was partially supported by the U.S. Department of Energy, U.S. Air Force including the Dual Use Science and Technology program and the Air Force Office of Scientific Research, and the New York State Research and Development Authority. Part of the work was also done in a Cooperative Research and Development Agreements with Los Alamos National Laboratory, Argonne National Laboratory and the Air Force Research Laboratory at Wright Patterson Air Force Base.

REFERENCES
[1]Y. Iijima, N. Tanabe, O. Kohno, and Y. Ikeno, "In-Plane Aligned YBa$_2$Cu$_3$O$_{7-x}$ Thin Films Deposited on Polycrystalline Metallic Substrates," *Appl. Phys Lett..* **60**, 770 (1992).

[2] S. R. Foltyn, P. N. Arendt, P. C. Dowden, R. F. DePaula, J. R. Groves, J. Y. Coulter, Q. Xia, M. P. Maley, and D. E. Peterson, "High-T$_c$ Coated Conductors – Performance of Meter-Long YBCO/IBAD Flexible Tapes, *IEEE Trans. Appl. Superconductivity* **9**, 1519 (1999).

[3] A. Usokin, J. Knoke, F. Garcia-Moreno, A. Issaev, J. Dzick, S. Sievers, and H. C. Freyhardt, "Large-area YBCO-Coated Stainless Steel Tapes with High Critical Currents," *IEEE Trans. Appl. Supercond.* **9**, 2248 (1999).

[4] D.P. Norton, A. Goyal, J.D. Budai, D. K. Christen, D. M. roeger, E. D. Specht, Q. He, B. Saffain, M. Paranthaman, C. E. Klabunde, D. F. Lee, B. C. Sales, F. A. List, "Epitaxial YBa$_2$Cu$_3$O$_7$ on Biaxially Textured (001) Ni : An Approach to High Critical Current Density Superconducting Tapes," *Science*, **274**, 755 (1996).

[5] Y. Sato, K. Matsuo, Y. Takahashi, K. Muranaka, K. Fujino, S. Hahakura, K. Ohmatsu, H. Takei, "Development of YBa$_2$Cu$_3$O$_Y$ Tape by Using Inclined Substrate Method," *IEEE Trans. Appl. Superconductivity* **11**, 3365 (2001).

[6] S. R. Foltyn, P. C. Dowden, P. N. Arendt, and J. Y. Coulter, "Coated Conductor Tape Development," 2000 U.S. Dept. of Energy Annual Peer Review, Washington D.C., July 17-19, 2000.

[7] P. Chou, Q. Zhong, Q. L. Li, K. Abazajian, A. Ignatiev, C. Y. Wang, E. E. Deal, and J. G. Chen, "Optimization of Jc of YBCO Thin Films Prepared by Photo-assisted MOCVD through Statistical Robust Design," *Physica C* **254**, 93 (1995).

[8] Y. Qiao, Y. Li, S. Sathiaraju, J. Reeves, K. Lenseth, and V. Selvamanickam, "An Overview of the Coated Conductor Progress at IGC-SuperPower," *Physica C* **382**, 48 (2002).

[9] V. Selvamanickam, H.-G.Lee, Y. Li, J. Reeves, Y. Qiao, Y.Y. Xie, K. Lenseth, G. Carota, M. Funk, K. Zdun, J. Xie, K. Likes, M. Jones, L. Hope, and D. W. Hazelton, "Scale up of High-Performance Y-Ba-Cu-O Coated Conductors," *Proc. Appl. Supercond. Conf.* Houston, Aug. 3 – 8, 2002 (to be published).

[10] V. Selvamanickam, H. G. Lee, Y. Li, X. Xiong, Y. Qiao, J. Reeves, Y. Xie, A. Knoll, and K. Lenseth, "Fabrication of 100 A Class, 1 m long Coated Conductor Tapes by Metal Organic Chemical Vapor Deposition and Pulsed Laser Deposition," *Physica C.* (*Proc. ISS*, Yokohama, Nov. 11-13, 2002) (in print)

[11] V. Selvamanickam, G. Carota, M. Funk, N. Vo, and P. Haldar, U. Balachandran, and M. Chudzik, P. Arendt, J. R. Groves, R. DePaula, B. Newnam, "High-Current Y-Ba-Cu-O Coated Conductor using Metal Organic Chemical-Vapor Deposition and Ion-Beam-Assisted Deposition," *IEEE Trans. Appl. Supercond.* **11**, 3379 (2000).

[12] A. Usokin, H. C. Freyhardt, A. Issaev, J. Dzick, J. Knoke, M. P. Oomen, H. –W. Neumueller, "Large-area YBCO-Coated Stainless Steel Tapes with High Critical Currents," *Proc. Appl. Supercond. Conf.* Houston, Aug. 3 – 8, 2002 (to be published).

Fabrication of High Temperature Superconductor

INCLINED-SUBSTRATE PULSED LASER DEPOSITION OF YTTRIA-STABILIZED ZIRCONIA TEMPLATE FILM FOR YBCO COATED CONDUCTORS

B. Ma, M. Li, B. L. Fisher, R. E. Koritala, R. M. Baurceanu, S. E. Dorris, and U. Balachandran
Energy Technology Division
Argonne National Laboratory
Argonne, IL 60439

ABSTRACT

Biaxially textured films of yttria-stabilized zirconia (YSZ) were deposited on Hastelloy C276 (HC) substrates by inclined-substrate pulsed laser deposition (ISPLD). This method is promising for fabrication of YSZ templates on poly-crystalline metallic tapes for coated conductor applications. Scanning electron microscopy showed columnar grains in ISPLD-YSZ films. X-ray pole figure analysis revealed good biaxial alignment in these films. The in-plane texture was determined to be $\approx 16°$ from the full-width at half maximum (FWHM) in the YSZ (111) ϕ-scan; and the out-of-plane texture was $\approx 8°$ from the FWHM in the YSZ (002) ω-scan. Before the deposition of YBCO films by pulsed laser deposition, a thin layer of CeO_2 was deposited on the ISPLD-YSZ. The YBCO deposited on ISPLD-YSZ-buffered HC substrates were biaxially textured. $T_c = 90$ K and $J_c = 180$ kA/cm^2 at 77 K in self-field were measured.

INTRODUCTION

$YBa_2Cu_3O_{7-\delta}$ (YBCO) coated conductors are promising for high-current carrying wires and other electric power devices operating at temperatures that approach liquid nitrogen [1-3]. Textured template films or buffer layers are needed for deposition of biaxially aligned YBCO films to overcome weak links at the grain boundaries and, therefore, to achieve high critical current density (J_c) in the YBCO films on metallic substrates [4]. Several techniques, including ion-beam-assisted deposition (IBAD), rolling-assisted biaxially textured substrates (RABiTS), and inclined-substrate deposition (ISD), have been developed in recent years [5-9].

We grew biaxially textured YSZ thin films on mechanically polished Hastelloy C276 (HC) substrates by inclined-substrate pulsed laser deposition (ISPLD). Ceria cap layers and YBCO films were subsequently deposited on ISPLD-YSZ-buffered metallic substrates by pulsed laser deposition (PLD). X-ray pole figures, ϕ-scan and ω-scan were used to analyze texture. Scanning electron microscopy (SEM) and atomic force microscopy (AFM) were utilized to study morphology and surface roughness. In this paper, we discuss the growth conditions, microstructure, and crystalline texture of ISPLD YSZ buffer layers and YBCO films deposited on polished HC substrates.

EXPERIMENTAL PROCEDURE

HC coupons ($\approx$5 mm wide and 10 mm long) were mechanically polished to a mirror finish with 0.25-μm diamond paste for use as substrates. Surface roughness of $\approx$3 nm was measured by AFM. A schematic illustration of the experimental setup is shown in Fig. 1. The polished substrate was mounted on a tiltable heater stage using silver paste. The substrate inclination angle (α), substrate normal with respect to the laser plume axle, was set to a desired value (45-55°). A Lambda Physik LPX 210i excimer laser, with a Kr-F$_2$ gas premixture as the lasing medium, was used for ablation of targets (Superconductive Components, 99.99% pure). YSZ, CcO$_2$ and YBCO targets are 45 mm in diameter and 6 mm thick. The size of laser spot focused at the rotating target was $\approx$12 mm^2, which produced an energy density of $\approx$2.0 J/cm^2. The distance between the target and the substrates was 4-7 cm. The desired oxygen partial pressure was obtained by flowing ultra-high-purity oxygen through the chamber. Details of deposition conditions are listed in Table 1. Layers of CeO$_2$ and YBCO were subsequently deposited by PLD on the ISPLD-YSZ buffered HC substrate at zero degree inclination angle.

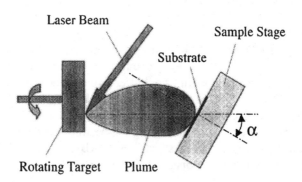

FIGURE 1. Schematic illustration of experimental setup for ISPLD system.

Fabrication of High Temperature Superconductor

Table 1. PLD Conditions for Growth of Biaxially Textured YSZ Films

Laser System	Lambda Physik LPX 210i
Laser Wavelength	248 nm (KrF)
Pulse Duration	25 ns
Energy Density	1-3 J/cm^2
Repetition Rate	20-90 Hz
Substrate Temperature	400-650°C
Inclination Angle	45-55°
Oxygen Partial Pressure	1-200 mTorr
Target-to-Substrate Distance	4-7 cm

Crystalline texture was measured by X-ray diffraction pole-figure analysis using Cu-K$_\alpha$ radiation. In-plane texture was characterized by the FWHM of ϕ-scans for the YSZ (111) reflection, and out-of-plane texture was characterized by the FWHM of ω-scans for YSZ (002). Surface morphology was investigated by SEM using a Hitachi S-4700-II. Surface roughness was measured by AFM using a Digital Instruments D3100 scanning probe microscope (operated in tapping mode). The superconducting critical transition temperature (T$_c$) and J$_c$ for the YBCO films were determined by the inductive method reported earlier [9].

RESULTS AND DISCUSSION

The X-ray diffraction 2θ scan patterns for the ISPLD YSZ films deposited at different temperature are shown in Fig. 2. Data indicate that out-of-plane orientation

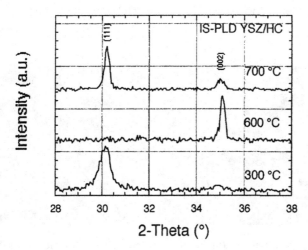

FIGURE 2. X-ray diffraction 2θ scan patterns for YSZ films deposited on HC substrate at different temperatures using ISPLD.

is dependent on deposition temperature. C-axis oriented films were obtained with a deposition temperature of ≈600°C. The (111) oriented YSZ films, whose texture was not in favor for the coated conductor applications, were obtained at higher (above 650°C) or lower (below 400°C) deposition temperatures. We will limit our discussion to c-axis oriented films in this paper.

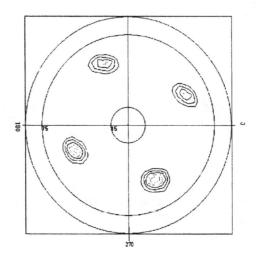

FIGURE 3. YSZ (111) pole figure for ISPLD YSZ film deposited at 600°C.

Figure 3 shows a typical X-ray diffraction pole figure of YSZ (111) for an ISPLD-YSZ film deposited at 600°C on an HC substrate. The four distinct poles indicate that the ISPLD-YSZ film was biaxially textured. In-plane texture measured

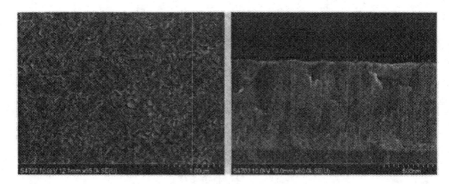

FIGURE 4. (a) Plan-view and (b) fracture cross-sectional view SEM images of ISPLD YSZ film deposited at 600°C.

Fabrication of High Temperature Superconductor

from the FWHM of the YSZ (111) φ-scan was ≈16.0°. The c-axis of ISPLD-YSZ film was slightly tilted, with a tilt angle of ≈6°, similar to that reported by Hasegawa et al. [10]. Out-of-plane texture measured from the FWHM of the YSZ (002) ω-scan was ≈7.8°. Gaps between grains were observed from the plan-view SEM image (Fig. 4a) of an ISPLD YSZ film. Columnar grains were observed from the cross-sectional fracture surface (Fig. 4b). This funding suggests that the texturing mechanism for ISPLD-YSZ films is most likely due to grain self-shadowing effect, similar to that for the IDS MgO films [11,12]. However, unlike the ISD MgO, textured YSZ films can hardly be grown by simple inclination of substrates and using electron beam evaporation, where the momentum energy of the atoms in YSZ was not sufficent. In order to have atoms in YSZ to have a better arrangement at their sublattices, additional momentum other than thermal energy must be supplied to these atoms. In the case of ISPLD, the energy needed for preferred lattice arrangement in YSZ was provided by the pulsed laser plasma/plume. Root-mean-square (RMS) surface roughness of 16 nm was measured by tapping mode AFM on the as-deposited YSZ films.

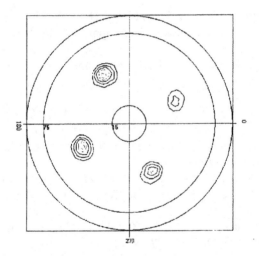

FIGURE 5. YBCO (103) pole figure of YBCO deposited on ISPLD YBZ buffered HC substrate.

YBCO deposited on ISPLD-YSZ buffered HC substrate was biaxially textured, as shown in the YBCO (103) pole figure (Fig. 5). A thin ceria layer was deposited prior to deposition of YBCO by PLD. The YBCO c-axis was parallel to the substrate normal, as illustrated by the four evenly distributed peaks in the YBCO (103) pole figure. FWHM = 18.2° was detected from the φ-scan of YBCO (103). T_c = 90 K and J_c = 180 kA/cm^2 at 77 K in self-field were measured.

CONCLUSIONS

Biaxially textured YSZ films were successfully grown by the ISPLD method. These films contained columnar grains. The surface roughness ≈16 nm was measured on the as-grown surface of an ISPLD YSZ thin films deposited on an HC substrate. In-plane texture measured by the FWHM in the YSZ (111) ϕ-scan was $\approx16°$. Out-of-plane texture measured by the FWHM in the YSZ (002) ω-scan was $\approx8°$. YBCO films deposited on ISPLD YSZ buffered HC substrates were biaxially textured with 18.2° FWHM in the ϕ-scan for YBCO (103). $T_c = 90$ K and $J_c = 180$ kA/cm^2 at 77 K in self-field were measured.

ACKNOWLEDGMENT

SEM analysis was performed in the Electron Microscopy Center for Materials Research at Argonne National Laboratory. This work was supported by the U.S. Department of Energy (DOE), Energy Efficiency and Renewable Energy, as part of a DOE program to develop electric power technology, under Contract W-31-109-Eng-38.

REFERENCES

1. D. K. Finnemore, K. E. Gray, M. P. Maley, D. O. Welch, D. K. Christen, and D. M. Kroeger, "Coated Conductor Development: An Assessment," *Physica C*, 320, 1-8 (1999).
2. Y. Iijima and K. Matsumoto, "High-Temperature-Superconductor Coated Conductors: Technical Progress in Japan," *Supercond. Sci. Technol.*, 13, 68-81 (2000).
3. J. O. Willis, P. N. Arendt, S. R. Foltyn, Q. X. Jia, J. R. Groves, R. F. DePaula, P. C. Dowden, E. J. Peterson, T. G. Holesinger, J. Y. Coulter, M. Ma, M. P. Maley, and D. E. Peterson, "Advance in YBCO-Coated Conductor Technology," *Physica C*, 335, 73 (2000).
4. D. Dimos, P. Chaudhari, and J. Mannhart, "Superconducting Transport Properties of Grain Boundaries in YBa$_2$Cu$_3$O$_7$ Bicrystals," *Phys. Rev. B*, 41, 4038-4049 (1990).
5. Y. Iijima, M. Kimura, T. Saitoh, and K. Takeda, "Development of Y-123-Coated Conductors by IBAD Process," *Physica C*, 335, 15 (2000).
6. C. P. Wang, K. B. Do, M. R. Beasley, T. H. Geballe, and R. H. Hammond, "Deposition of In-Plane Textured MgO on Amorphous Si$_3$N$_4$ Substrate by Ion-Beam-Assisted Deposition and Comparisons with Ion-Beam-Assisted Deposited Yttria-Stabilized-Zirconia," *Appl. Phys. Lett.*, 71, 2955-2958 (1997).
7. A. Goyal, D. P. Norton, J. D. Budai, M. Pranthaman, E. D. Specht, D. M. Kroeger, D. K. Christen, Q. He, B. Saffian, F. A. List, D. F. Lee, P. M. Martin, C. E. Klabunde, E. Hardtfield, and V. K. Sikka, "High Critical Current Density Superconducting Tapes by Epitaxial Deposition of YBCO Films on Biaxially Textured Metals," *Appl. Phys. Lett.*, 69, 1975 (1996).
8. M. Bauer, R. Semerad, and H. Kinder, "YBCO Films on Metal Substrates with Biaxially Aligned MgO Buffer Layers," *IEEE Trans. Appl. Supercond.*, 9, 1502, (1999).

9. B. Ma, M. Li, Y. A. Jee, B. L. Fisher, and U. Balachandran, "Inclined Substrate Deposition of Biaxially Textured Magnesium Oxide Films for YBCO Coated Conductors," *Physica C*, 366, 270-276 (2002).

10. K. Hasegawa, K. Fujino, H. Mukai, M. Konishi, K. Hayashi, K. Sato, S. Honjo, Y. Sato, H. Ishii, and Y. Iwata, "Biaxially Aligned YBCO Film Tapes Fabricated by All Pulsed Laser Deposition," *Appl. Supercond.*, 4, 487-493 (1996).

11. B. Ma, M. Li, R. E. Koritala, B. L. Fisher, A. R. Markowitz. R. A. Erck, R. Baurceanu, S. E. Dorris, D. J. Miller, and U. Balachandran, "Pulsed Laser Deposition of YBCO Films on ISD MgO Buffered Metal Tapes," *Supercond. Sci. Technol.*, 16, 464-472 (2003).

12. B. Ma, M. Li, B. L. Fisher, R. E. Koritala, and U. Balachandran, "Inclined Substrate Deposition of Biaxially Aligned Template Films for YBCO Coated Conductors," *Physica C*, 382, 38-42 (2002).

EVALUATING SUPERCONDUCTING YBCO FILM PROPERTIES USING X-RAY PHOTOELECTRON SPECTROSCOPY

Paul N. Barnes, Justin C. Tolliver, and Timothy J. Haugan
Air Force Research Laboratory
AFRL/PRPG, Building 450
Wright-Patterson AFB, OH 45433

Sharmila M. Mukhopadhyay
Wright State University
Russ Engineering Center
Dayton, OH 45435

John T. Grant
University of Dayton
300 College Park
Dayton, OH 45469

ABSTRACT

Initial results have been recently reported that suggest a potential correlation exists between the full-width-half-maximum (FWHM) of the Y(3d) peak obtained by x-ray photoelectron spectroscopy (XPS) and the critical current density a $YBa_2Cu_3O_{7-x}$ film can carry. In particular, the Y($3d_{5/2}$) demonstrated a stronger correlation. Transport currents were determined by the 4-point contact method using the 1 μV/cm criterion. An apparent correlation was also suggested between the Y(3d) FWHM and ac loss data from magnetic susceptibility measurements. In this report, a few additional data points were acquired to further test the usefulness of the correlations. Samples were created by pulsed laser deposition of $YBa_2Cu_3O_{7-x}$ on $LaAlO_3$ substrates.

INTRODUCTION

Significant progress has been made in the development of the high temperature superconducting (HTS) $YBa_2Cu_3O_{7-x}$ (YBCO) coated conductors.[1-4] These achievements have been accomplished by using a variety of techniques investigating multiple aspects of the conductor architecture and growth.[5-6] Successful long length development of the YBCO coated conductor will result in its availability for use in a variety of commercial applications such as power transmission cables, high field magnets, transformers, high power generators and motors, etc.

However, many improvements can still be made in the properties of the coated conductor as well as production of longer lengths.[7-9] Continuing development of the coated conductor necessitates not only a detailed study of deposition techniques, source materials, deposition conditions, and substrates employed, but also characterization of the final deposited films. A technique used in this paper is x-ray photoelectron spectroscopy (XPS), especially of the YBCO layer. XPS was used to investigate the chemical and microstructural profiles of YBCO coated conductor samples.[10-11] Use of the technique can give consideration to interfacial issues and determine composition and chemistry at different depths in the conductor architecture.

BACKGROUND

Comparison of the XPS spectra for the various constituents of YBCO coated conductor samples revealed that the Y(3d) photoelectron peak shape observed from the YBCO layer differed among various samples investigated.[10-11] This difference in XPS peak shape may indicate that there could be some difference in the atomic co-ordination between the samples. The observed difference, when compared to the particular film's properties, may also indicate the possibility of a correlation with the particular quality of the film.

In a previous study by Beyer et al.,[12] a difference in samples was also noted that depended on the particular substrate used. In this study, two series of samples were prepared by hollow cathode sputtering of the YBCO film, between 100 nm to 300 nm thick. One series was grown on (100) MgO and the other on (001) YSZ buffered r-plane sapphire. The T_cs were typically between 80-91 K, a wide variance by recent deposition standards. During the various treatments, the Y emission remained fairly stable, but the Ba ($3d_{5/2}$) structure showed distinct differences in its lineshape in the YBCO deposited on the two different substrates. Beyer at al. noted that the differences were not caused by surface impurities but were characteristic of the samples. A structural deviation was assumed to be responsible for the difference. Even so, this does not rule out that the differences might correlate to the film's current transport properties since some substrates indeed provide better subsequent films than others.

Another possible correlation cited in the previous work is the relationship of the Y(3d) peak FWHM and the samples ac loss data from magnetic susceptibility measurements. It has been previously observed that with widening of the temperature-dependent ac susceptibility curves with increasing applied magnetic field the quality of the YBCO film generally decreases.[13] However, a documented study of this correlation to current transport properties using the loss component of ac susceptibility data has not been published, making it unclear how effective this correlation is.[14] Even so, the relationship between the Y(3d) peak FWHM and the ΔT of the magnetic field lines reinforces the possibility. In this case, ΔT of the magnetic field lines refers to either the difference between the temperatures at which the maxima in the ac loss occurs for the 0.025 and 2.2 Oe applied magnetic

Fabrication of High Temperature Superconducto▶

fields (peak to peak of χ'') or to the FWHM of the 2.2 Oe (in particular) magnetic field data χ'' vs. T

EXPERIMENTAL

In this study, samples were created by pulsed laser deposition of YBCO on single crystal $LaAlO_3$. Previously, most samples were made in the same manner although other samples are included such as YBCO on buffered metallic tape and buffered single crystals.[10] The laser ablation was accomplished using a Lambda Physik LPX 305i excimer laser at the KrF transition, $\lambda = 248$ nm with a 25 ns pulse width. Mounting of the substrates in the deposition chamber was accomplished using silver paste. Specific details of the deposition conditions were given previously.[10] The YBCO layers of the newly created samples were between 0.25 to 0.40 micrometers thick.

Since the previous study contained data representing a limited number of low quality samples, i.e. $J_c < 10^6$ A/cm^2, the three samples used here were particularly chosen for this study due to their poor film quality. One sample was unintentionally made with low quality and the other two were intentionally made so by specifically reducing the deposition temperature in the PLD chamber by ~100 °C. The critical transition temperature (T_c) of YBCO was measured by ac magnetic susceptibility and the critical current density (J_c) by four-contact transport current measurement.

The composition and chemistry of each sample was measured by X-ray photoelectron spectroscopy (XPS) using a Kratos AXIS Ultra. The monochromatic Al K_α x-ray line was used for enhanced spectral resolution. The analysis spot size was approximately 110 μm^2. An electron flood charge neutralizer was used during analysis to avoid charge build-up differences between different surfaces (if any). Ion beam sputtering was performed using a mini-beam ion gun. Ar^+ ions were used at an energy of 3 keV. In the raster setting used, the sputtered area was approximately a 1 mm x 0.5 mm elliptical region, several times larger than the spot size analyzed by XPS. Spectroscopic analysis was performed on as received surfaces and after subsequent sputtering to avoid the influence of surface contamination or any chemical reactivity with air.

RESULTS AND DISCUSSION

The J_cs of the samples were 1.5 x 10^5 A/cm^2 for the unintentionally made poor sample, 2 x 10^5 A/cm^2 for the one low deposition temperature sample and < 10^4 A/cm^2 for the other low deposition temperature sample. All three YBCO samples in this study had a reduced T_c. Refer to Figure 1 for the low temperature deposition sample with the higher critical current density. The T_c for the sample made with the normal deposition temperature had a T_c of ~86.1 K. Since samples were exposed to air the top surface was removed by sputtering. The before and after sputtering values obtained by XPS are largely uncorrelated as evidenced by

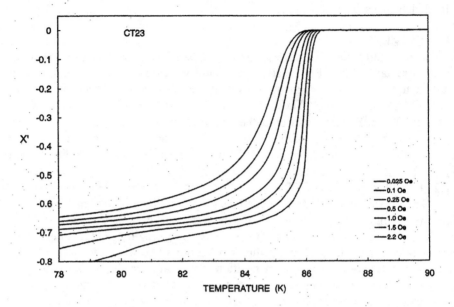

Fig. 1. The ac susceptibility data for sample CT23 made with the low deposition temperature but had a nominal J_c of 2×10^5 A/cm^2. The different curves result from the different applied fields listed in the legend—the field increases from right to left.

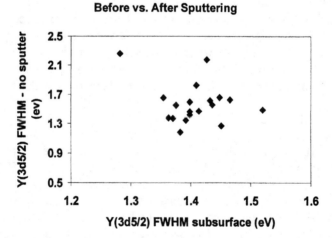

Fig. 2. Comparison of XPS data taken before and after sputtering of the surface for contaminate removal.

the scatter of the data given in Figure 2 which shows values for all samples included in the previous study, although not reported there.

In Figure 3, the relative peak intensity ratios of all samples for both studies are shown. The two stray points depicted were the two that were intentionally made poor by lowering the deposition temperature. Since all other samples had similar peak intensity ratios, it is likely that the difference caused by the low deposition temperature is due to inclusions and alternate phases incorporated during

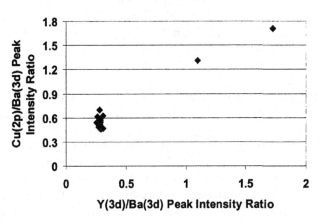

Fig. 3. Comparison of the relative XPS peak intensities of the various samples. The two compositionally stray points were made with the low deposition temperature.

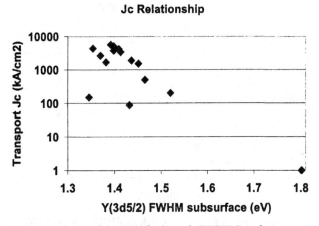

Fig. 4. Comparison of the Y(3d$_{5/2}$) peak FWHM and transport J$_c$.

deposition. As such, these compositionally stray points will be excluded from the correlation charts as being invalid for comparison.

In Figure 4 the correlation of the FWHM for the Y($3d_{5/2}$) peak and the critical transport current density (four point resistive) is given. The new point provided by the additional sample is located at 1.35 eV FWHM Y($3d_{5/2}$) with 0.15 MA/cm^2. For the given plot, this data resulted in a stray point from the other points which indicate a correlation. The point located at 1.8 eV is not stray since it fits the pattern of low J_c correlated to a larger Y($3d_{5/2}$) FWHM. It is not clear if the lower transport current was a result of cracks in the film which can cause a lower transport current and yet not affect the XPS spectra. However, it is interesting to note that this sample also resulted in a stray point to a greater degree on the following figure, Figure 5, which gives the relationship between the FWHM for the Y($3d_{5/2}$) peak and the FWHM of the critical transition temperature by susceptibility measurement.

Figure 5 depicts the relationship between the FWHM for the Y($3d_{5/2}$) peak and the FWHM of the 2.2 Oe magnetic field line at the critical transition temperature by ac susceptibility measurement. As previously mentioned, the new data point on this plot (5 K for ΔT) also strayed from the others based on the expected relationship. Exactly why there is a corroborative stray form the original data in both graphs is not clear, but a correlation between the points apparently still exists. This clearly indicates the need for additional data to determine the effectiveness of the correlations as well as determining the exact relationship of the two plots presented here. Figure 6 shows the χ'' susceptibility data for sample CT23, from Figure 1, from which the FWHM for the Y(3d5/2) peak was derived. The FWHM of the 2.2 Oe magnetic field data was used, although this does not imply that this is the predominant usage.[13-14]

Tc FWHM Relationship

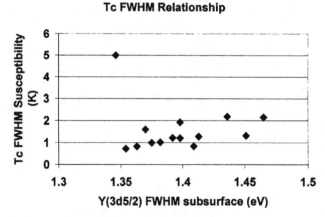

Fig. 5. Comparison of the XPS Y($3d_{5/2}$) FWHM and the T_c FWHM determined by ac susceptibility measurement (2.2 Oe).

Fabrication of High Temperature Superconducto

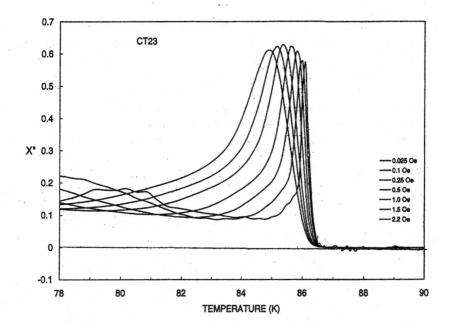

Fig. 6. χ'' versus temperature plot of the ac susceptibility data. 0.025 to 2.2 Oe magnetic field used. The different curves result from the different applied fields listed in the legend—the field increases from right to left.

CONCLUSION

It is seen that the FWHM of certain XPS peaks as well as XPS cationic peak ratios averaged over the analysis area can vary between YBCO samples. A possible correlation of the FWHM of the $Y(3d_{5/2})$ XPS peak of YBCO to the thin film quality, specifically the critical transport current density (J_c) may exist but is not clear. This relationship is dependent upon the appropriately phased YBCO for a proper comparison and YBCO samples whose nonstoichiometric composition will lead to an alternate XPS spectra. Broadening of the FWHM of the $Y(3d_{5/2})$ XPS peak can indicate alternate undesirable bonding in the YBCO. More data is necessary to fully verify these relationships.

ACKNOWLEDGMENT

The authors gratefully acknowledge the assistance of Lyle Brunke, Julianna Evans, and Timothy Campbell of the Air Force Research Laboratory for their assistance in film preparation, x-ray diagnostics, and making the critical current transport measurements.

REFERENCES

[1] V. Selvamanickam, H.G. Lee, Y. Li, X. Xiong, Y. Qiao, J. Reeves, Y. Xie, A. Knoll, and K. Lenseth, "Fabrication of 100 A class, 1 m long coated conductor tapes by metal organic chemical vapor deposition and pulsed laser deposition," Physica C, **392-396** [2], pp. 859-862 (2003).

[2] D.T. Verebelyi, U. Schoop, C. Thieme, X. Li, W. Zhang, T. Kodenkandath, A.P. Malozemoff, N. Nguyen, E. Siegal, D. Buczek, J. Lynch, J. Scudiere, M. Rupich, A. Goyal, E.D. Specht, P. Martin, and M. Paranthaman, "Uniform performance of continuously processed MOD-YBCO-coated conductors using a textured Ni–W substrate," Superconductivity Science & Technology, **16**, pp. L19-L22 (2003).

[3] J.R. Groves, P.N. Arendt, S.R. Foltyn, Q. Jia, T.G. Holesinger, H. Kung, R.F. DePaula, P.C. Dowden, E.J. Peterson, L. Stan, and L.A. Emmert, "Recent progress in continuously processed IBAD MgO template meters for HTS applications," Physica C, **382** [1], pp. 43-47 (2002).

[4] A. Goyal, D.F. Lee, F.A. List, E.D. Specht, R. Feenstra, M. Paranthaman, X. Cui, S.W. Lu, P.M. Martin, D.M. Kroeger, D.K. Christen, B.W. Kang, D.P. Norton, C. Park, D.T. Verebelyi, J.R. Thompson, R.K. Williams, T. Aytug, and C. Cantoni, "Recent progress in the fabrication of high-J_c tapes by epitaxial deposition of YBCO on RABiTS," Physica C, **357-360** [2], pp. 903-913 (2001).

[5] Air Force Office of Scientific Research High Temperature Superconducting Coated Conductor Review, St. Petersburg, Florida, January 22-24, 2003.

[6] Department of Energy Superconductivity for Electric Systems Annual Peer Review, Washington, DC, July 23-25, 2003.

[7] C.B. Cobb, P.N. Barnes, T.J. Haugan, J. Tolliver, E. Lee, M. Sumption, E. Collings, C.E. Oberly, "Hysteresis loss reduction in striated YBCO," Physica C., **382**, pp. 52-56 (2002).

[8] T.J. Haugan,, P.N. Barnes, I. Maartense, E.J. Lee, M. Sumption, and C.B. Cobb, "Island growth of Y_2BaCuO_5 nanoparticles in $(211_{\sim1.5nm}/123_{\sim10nm})xN$ composite multilayer structures to enhance flux pinning of $YBa_2Cu_3O_{7-d}$ films," J. Mater. Res., **18**, pp.2618-2623 (2003).

[9] T.J. Haugan, M.E. Fowler, J.C. Tolliver, P.N. Barnes, W. Wong-Ng, and L.P. Cook, "Flux Pinning and Properties of Solid-Solution $(Y,Nd)_{1+x}Ba_{2-x}Cu_3O_{7-d}$ Superconductors," Ceramic Trans., **140**, pp. 299-325 (2003).

[10] P.N. Barnes, S. M. Mukhopadhyay, S. Krishnaswami, T.J. Haugan, J.C. Tolliver, and I. Maartense, "Correlation between the XPS peak shapes of $Y_1Ba_2Cu_3O_{7-x}$ and film quality," IEEE Trans. on Appl. Superconductivity, **13**, pp. 3643-3646 (2003).

[11] P.N. Barnes, S. Mukhopadhyay, R. Nekkanti, T. Haugan, R. Biggers, and I. Maartense, "XPS depth profiling studies of YBCO layer on buffered substrates," Advances in Cryogenic Engineering, **48B**, pp. 614-618 (2002).

Fabrication of High Temperature Superconducto

[12]J. Beyer, Th. Schurig, S. Menkel, Z. Quan, and H. Koch, "XPS investigation of the surface composition of sputtered YBCO thin films," Physica C, **246**, pp.156-162 (1995).

[13]I. Maartense and A. K. Sarkar, "Annealing of pressure-induced structural damage in superconducting Bi-Pb-Sr-Ca-Cu-O ceramic," J. Mater. Res., **8**, pp. 2177 (1993).

[14]P. N. Barnes, T.J. Haugan, S. Sathiraju, I. Maartense, A.L. Westerfield, R.M. Nekkanti, L.B. Brunke, T.L. Peterson, J.M. Evans, and J.C. Tolliver, "Correlation of AC Loss Data from Magnetic Susceptibility Measurements with YBCO Film Quality," to be presented at MRS Fall Meeting, Boston, MA, December 1-5, 2003.

Fabrication of High Temperature Superconducto

DEVELOPMENT OF LOW-COST ALTERNATIVE BUFFER LAYER ARCHITECTURES FOR YBCO COATED CONDUCTORS

M. Parans Paranthaman, T. Aytug, H.Y. Zhai, H.M. Christen, D.K. Christen, A. Goyal, L. Heatherly, and D.M. Kroeger
Oak Ridge National Laboratory, Oak Ridge, TN 37831, USA

ABSTRACT

We have developed a simpler alternative buffer layer architecture for Rolling-Assisted Biaxially Textured Substrates (RABiTS) approach. Cube textured MgO buffers were grown directly on biaxially textured Ni and Ni-W substrates. MgO has been proved to be a good diffusion barrier for oxygen. In addition, LaMnO$_3$ (LMO) has been identified as a compatible buffer layer for MgO and it also provides a good template for growing high current density YBCO films. We have optimized the growth of LMO layers on MgO/Ni using rf sputtering. Sputtered CeO$_2$ cap layers were also grown on LMO layers for compatibility with Ex-situ YBCO process. YBCO films with a J$_c$ of 500,000 A/cm^2 at 77 K and self-field were grown on this newly developed architecture of CeO$_2$/LMO/MgO/Ni using pulsed laser deposition.

INTRODUCTION

The Rolling-Assisted Biaxially Textured Substrates (RABiTS) and Ion-Beam Assisted Deposition (IBAD) approaches have been identified recently as the leading techniques to fabricate long lengths of high performance YBCO coated conductors.[1-4] In the standard RABiTS approach, a four-layer architecture of CeO$_2$/YSZ/Y$_2$O$_3$/Ni/Ni-W is used to fabricate long lengths of buffered tapes by the epitaxial deposition on the thermo mechanically-textured Ni alloy substrates. In an effort to develop low-cost alternative buffer layer architecture, we focused our studies on investigating both oxygen and metal diffusion layers. In order to develop a robust oxygen diffusion barrier layer, MgO has been chosen as the potential candidate since the oxygen diffusivity in MgO at 800 °C is 8 x 10^{-22} cm^2/sec. The oxygen diffusion into the metal/buffer interface could potentially oxidize the substrate surface to NiO, WO$_3$, etc. based on the substrate composition and thermodynamical considerations. This may lead to delamination of both buffers and superconductors. It could also adversely affect the mechanical properties of the conductors. To overcome these issues, MgO has been grown directly on textured Ni or Ni-W3% substrates for the first time. In the past, we have demonstrated the epitaxial growth of MgO layers on both Ag/Pd-buffered Ni substrates and Pd-buffered Ni substrates.[5] In addition, LaMnO$_3$ (LMO) has also been identified as a good diffusion barrier to nickel contamination.[6,7] The pseudo-cubic lattice parameter of LMO (3.88 Å) is closely matched to YBCO; the lattice mismatch is less than 0.8%. We have also shown recently that LMO is compatible with MgO surfaces.[8] Based on these studies, we have also grown LMO on MgO-buffered Ni substrates. Here we report our recent results obtained on MgO-RABiTS architecture.

EXPERIMENTAL PROCEDURE

The MgO seed layers were grown directly on biaxially textured Ni or Ni-W3% substrates using electron beam evaporation. As-rolled Ni or Ni-W tapes were cleaned by ultrasonification in iso-propanol. The tapes were then annealed in a high vacuum system at 1200-1300 °C in the presence Ar/H$_2$(4%) and H$_2$S gas atmospheres to obtain the desired cube texture with a complete surface coverage of sulfur c(2x2) superstructures. The substrates used were 1 cm wide and 50 μm thick. Biaxially textured Ni substrates were mounted on a heater in the e-beam system. After the vacuum in the chamber had reached a background pressure of 1 x 10^{-6} Torr at room temperature, the substrates were heated to various temperatures ranging from 300 – 600 °C. The MgO layers were deposited on the Ni substrates at an optimum temperature of 400 °C. The crucibles used were graphite. Magnesium Oxide crystals were used as the source material. The deposition rate for MgO was 0.5 nm/sec with the operating pressure of 10^{-5} Torr, and the final thickness was varied from 30 – 300 nm. We have deposited 60 nm thick LMO buffer layers on MgO-buffered Ni substrates by rf-magnetron sputtering. The oxide sputter targets were made from single-phase LMO powders, prepared by solid-state reaction, which were loosely packed in a 4" copper tray. Typical sputter conditions consisted of 2-5 x 10^{-5} Torr of H$_2$O with a total pressure of 3 mTorr forming gas (Ar/H$_2$ 4%). The water pressure is sufficient to oxidize the film to form stoichiometric LaMnO$_3$, when grown at a substrate temperature of 650-750 °C. The deposition rate was ~0.06 nm/sec. Using rf magnetron sputtering; ~ 20 nm-thick CeO$_2$ cap layers were also deposited on the LMO-buffered MgO/Ni substrates at 780 °C in 10 mTorr of Ar/H$_2$ (4%) gas and a water pressure of 2 x 10^{-5} torr. The plasma power was 75 W. YBCO was deposited by pulsed laser deposition at 790 °C in 120 m torr oxygen with an average laser energy of 400-410 mJ using a stoichiometric YBCO target, followed by annealing under 550 Torr oxygen during cool down. Typical YBCO thickness was 200 nm.

Fabrication of High Temperature Superconducto

The crystalline structure of the films was analyzed by X-ray diffraction techniques. SEM micrographs were taken using a Hitachi S-4100 field emission microscope. The thickness of both buffer layers and YBCO were determined by Rutherford Backscattering Spectroscopy. The films were then prepared for current density measurements by depositing silver for current and voltage leads, followed by oxygen annealing at 500 °C for 1 h. The transport critical current density, J_c, was measured using a standard four-point probe technique with a voltage criterion of 1 μV/cm.

RESULTS AND DISCUSSION

Typical θ-2θ scan for a 30 nm thick MgO film grown on textured Ni substrate is shown in Figure 1. These scans indicate the presence of c-axis aligned films. It is interesting to note that there is no NiO present in the film. Detailed

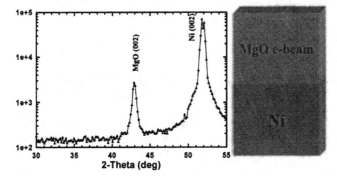

FIGURE 1. A typical θ-2θ scan for a 30 nm thick MgO film on textured Ni substrate. The MgO film has a preferred c-axis orientation.

XRD results from ω and φ scans (as shown in Figure 2) revealed good epitaxial texturing. The full width at half maximum (FWHM) values for Ni (002) and MgO (002) are 9.6° and 4.7°, and those of Ni (111) and MgO (220) are 8.0° and 7.4°, respectively. There was a significant improvement in the out-of-plane texture. This could be due to the smoothness of the MgO layers. As shown in Figure 3, the MgO (220) pole figure revealed the presence of a single four-fold cube texture. Similarly, highly aligned MgO layers were grown on both Ni-W3% and Ni/Ni-W3% substrates. SEM micrographs for both 30 nm and 300 nm thick MgO seeds are shown in Figure 4 (a) and (b). Sample morphology of 30 nm thick MgO layer is smooth, uniform, crack-free and dense. MgO has also excellent coverage at grain boundaries. However, 300 nm thick MgO layers were cracked in orthogonal regions. This could be due to either thermal expansion or lattice mismatch between Ni and MgO, causing the release of strain in MgO at higher thicknesses. Hence, it is essential to grow thin crack-free MgO. The AFM images obtained on both 60 nm and 300 nm thick MgO surfaces are shown in Figure 5. The surface roughness, Ra obtained on 60 nm thick MgO surface is 7 nm and that on 300 nm thick MgO surface is 5.8 nm.

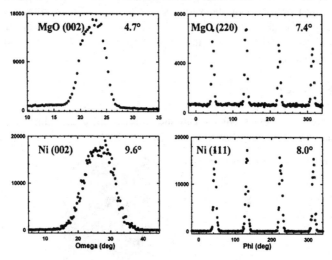

FIGURE 2. The ω and φ scans obtained for a 30 nm thick e-beam MgO film grown on textured Ni substrate. The FWHM values for each scan are shown inside the scans.

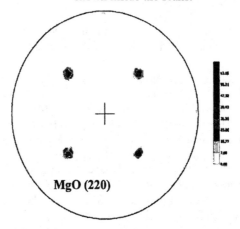

FIGURE 3. The typical MgO (220) pole figure obtained on a 30 nm thick e-beam MgO film grown on textured Ni substrate.

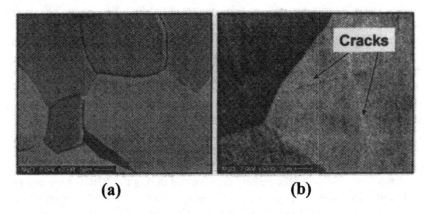

(a)   **(b)**

FIGURE 4. SEM micrograph obtained on (a) crack-free 30 nm thick and (b) cracked 300 nm thick MgO surface.

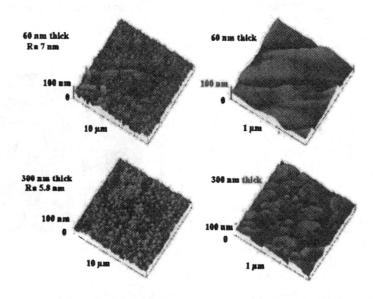

FIGURE 5. AFM images obtained on both 60 and 300 nm thick MgO surface.

Typical θ-2θ scan for a 60 nm thick sputtered LaMnO₃ film grown on e-beam MgO buffered Ni substrate is shown in Figure 6. These scans indicate the presence of a highly c-axis aligned LaMnO₃ film. Figure 7 shows detailed X-ray results obtained from ω and φ scans on LaMnO₃ layers. It revealed very good epitaxial texturing. The full width at half maximum (FWHM) values for Ni (002), MgO (002), and LMO (004) are 8.2°, 4.8°, and 5.2°, and those of Ni (111), MgO (220), and LMO (222) are 8.4°, 7.5°, and 7.2°, respectively. Typical θ-2θ scan for a 20 nm thick PLD-YBCO film grown on sputtered CeO₂ capped LMO/MgO/Ni substrate is shown in Figure 8. These scans indicate the presence of highly c-axis aligned CeO₂ and YBCO films. The total thickness of the buffer layers were ~ 100 nm. A J_c of 500,000 A/cm² at 77 K and self-field was obtained on these films. Some of the buffers delaminated during the YBCO growth. The main cause for the delamination is unknown at this time. Since there is no oxygen transport through MgO layers to the substrates during the YBCO growth, the formation of self-passivating layer of NiWO₄ may be limited. Recent cross-sectional TEM studies on high I_c thick YBCO film based RABiTS have shown that the presence of NiWO₄ layer may be beneficial.[9] Efforts are being made to optimize the YBCO growth conditions and also to understand the role of MgO seeds in detail.

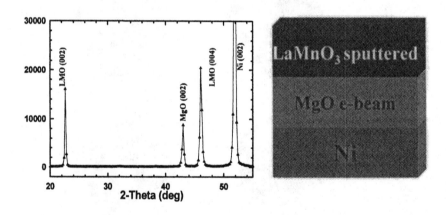

FIGURE 6. A typical θ-2θ scan for a 60 nm thick Sputtered LaMnO₃ film on e-beam MgO-buffered Ni substrate. The LaMnO₃ film has a preferred c-axis orientation.

Fabrication of High Temperature Superconducto

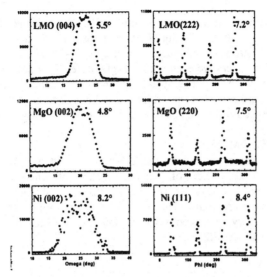

FIGURE 7. The ω and φ scans obtained for a sputtered 60 nm thick LaMnO₃ film on e-beam MgO buffered Ni substrate. The FWHM values for each scan are shown inside the scans.

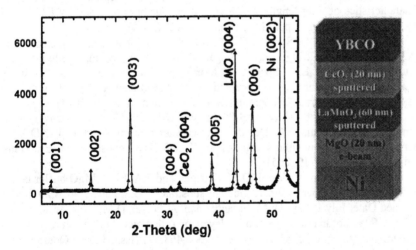

FIGURE 8. A typical θ-2θ scan for a 200 nm thick PLD-YBCO film on sputtered LaMnO₃-buffered MgO/Ni substrate. The YBCO film has a preferred c-axis orientation.

SUMMARY

We have demonstrated that MgO films can be grown epitaxially with a single cube-on-cube orientation on textured Ni substrates. The microstructure of the 30 nm thick e-beam grown MgO films was dense, crack-free and continuous. Highly aligned LaMnO₃ layers were grown on MgO-buffered Ni substrates. Sputtered CeO₂ cap layers were also developed for these architectures for compatibility with Ex-situ-YBCO process. Our preliminary results indicate that PLD-YBCO films with a J_c of 500,000 A/cm^2 can be obtained on CeO₂/LMO/MgO/Ni substrates.

ACKNOWLEDGEMENTS

This work was supported by the U.S. Department of Energy, Division of Materials Sciences, Office of Science, Office of Energy Efficiency and Renewable Energy, Office of Distributed Energy and Electric Reliability - Superconductivity Program. This research was performed at the Oak Ridge National Laboratory, managed by U.T.-Battelle, LLC for the USDOE under contract DE-AC05-00OR22725.

REFERENCES

[1] A. Usoskin, H.C. Freyhardt, A. Issaev, J. Dzick, J. Knoke, M.P. Oomen, M. Leghissa, and H-W. Neumueller, "Large area YBCO-coated stainless steel tapes with high critical currents," IEEE Trans on Appl. Supercond. **13**, 2452 (2003).

[2] Y. Iijima, K. Kakimoto, and T. Saitoh, "Fabrication and transport characteristics of long length Y-123 coated conductors processed by IBAD and PLD," IEEE Trans. on Appl. Supercond. **13**, 2466 (2003).

[3] D.T. Verebelyi, U. Schoop, C. Thieme, X. Li, W. Zhang, T. Kodenkandath, A.P. Malozemoff, N. Nguyen, E. Siegal, D. Buczek, J. Lynch, J. Scudiere, M. Rupich, A. Goyal, E.D. Specht, P. Martin, and M. Paranthaman, "Uniform performance of continuously processed MOD-YBCO-coated conductors using a textured Ni-W substrate," Supercond. Sci. and Technol. **16**, L19 (2003).

[4] J.R. Groves, P.N. Arendt, S.R. Foltyn, Q.X. Jia, T.G. Holesinger, L.A. Emmert, R.F. DePaula, P.C. Dowden, and L. Stan, "Improvement of IBAD MgO template layers on metallic substrates for YBCO HTS deposition," IEEE Trans. on Appl. Supercond. **13**, 2651 (2003).

[5] M. Paranthaman, A. Goyal, D.M. Kroeger, and F.A. List, "MgO buffer layers on rolled nickel or copper as superconductor substrates," U.S. Patent No. 6,261,704 (Issued Date: July 17, 2001); M. Paranthaman, A. Goyal, D.M. Kroeger, and F.A. List, "Method of making MgO buffer layers on rolled nickel or copper as superconductor substrates," U.S. Patent No. 6,468,591 (Issued Date: October 22, 2002).

[6] M.P. Paranthaman, T. Aytug, H.Y. Zhai, S. Sathyamurthy, A. Goyal, P.M. Martin, D.K. Christen, R.E. Erickson, and C.L. Thomas, "Single Buffer Layer

Fabrication of High Temperature Superconductor

Technology for YBCO Coated Conductors," Mater. Res. Soc. Symp. Proc., Vol. **689**, 323 (2002)

[7]T. Aytug, M. Paranthaman, H.Y. Zhai, H.M. Christen, S. Sathyamurthy, D.K. Christen, and R.E. Ericson, "Single buffer layers of $LaMnO_3$ or $La_{0.7}Sr_{0.3}MnO_3$ for the development of $YBa_2Cu_3O_{7-\delta}$-coated conductors: A comparative study," J. Mater. Res. **17**, 2193 (2002).

[8]M. Paranthaman, T. Aytug, D.K. Christen, P.N. Arendt, S.R. Foltyn, J.R. Groves, L. Stan, R.F. DePaula, H. Wang, and T.G. Holesinger, "Growth of thick $YBa_2Cu_3O_{7-\delta}$ films carrying a critical current of over 230 A/cm on single $LaMnO_3$-buffered ion-beam assisted deposition MgO substrates," J. Mater. Res. **18**, 2055 (2003).

[9]S. Kang, A. Goyal, K. J. Leonard, N. Rutter, D. F. Lee, D. M. Kroeger and M. Paranthaman, "High Critical Current $YBa_2Cu_3O_{7-\delta}$ Thick Films on Rolling-Assisted Biaxially Textured Substrates (RABiTS)," *J. Mater. Res.* (in press); K.J. Leonard, S. Kang, A. Goyal, K.A. Yarborough, and D.M. Kroeger, "Microstructural characterization of thick $YBa_2Cu_3O_{7-\delta}$ films on improved rolling-assisted biaxially textured substrates," J. Mater. Res. **18**, 1723 (2003).

IMPROVEMENT OF THE TEXTURE IN AG SUBSTRATES FOR HIGH TEMPERATURE SUPERCONDUCTOR DEPOSITION

D.M.Liu, M.L.Zhou, E.D.Li, W.Liu, Y.C.Hu, B.Zong, M.Liu and T.Y.Zuo
The Key Laboratory of Advanced Functional Materials, Ministry of Education, China;
College of Material Sciences and Engineering, Beijing University of Technology, Beijing 100022, P.R.China

ABSTRACT

By means of cold rolling control, purity control and annealing control a textured Ag tape with very strong {110}<110> component for HTS deposition has been obtained. It was found that, the oxygen content in silver might not exceed 25ppm, cold-rolling reduction should be higher than 92%, annealing temperature may be at about 900°C and the annealing treatment should be carried out in gaseous atmospheres.

INTRODUCTION

Several research groups reported that YBCO film possessing high Jc can be deposited on textured Ag substrate [1-7]. It is generally accepted that the {110}<110> texture is suitable for the growth of biaxially textured YBCO films. Some investigations have involved obtaining such kind of textured Ag foil [8~10]. However, the Jc of YBCO film deposited on it is not as good as that deposited on other substrates. One of the main problems of Ag substrate is that the texture in Ag is not sharp enough. Usually there exist not a single texture component and some scattered orientations around the main orientation, after deposition the texture in YBCO film can also have a multi-component texture. In order to obtain a sharp one-component texture in Ag tape, it is necessary to improve the forming conditions of such a texture carefully. For a {110}<110> oriented texture, many key factors have been confirmed [8-10]. However, some of them, such as oxygen content, are only qualitative description. Others of them, such as annealing atmosphere, are doubtful. Therefore, these key factors should be studied in detail.

In this work, the influence of a series of processing parameters on the recrystallization texture have been studied, including the oxygen content in the Ag ingot, the rolling reduction, and the annealing temperature and atmosphere. A reproducible method to prepare sharp {110}<110> textured tapes was put forward.

1. EXPERIMENTS

99.99wt% commercial silver was used as starting material, the oxygen content is 170ppm. The silver was melted in a induction furnace in vacuum of $5*10^{-2}$ Pa at 1050°C to reduce the oxygen content. Two ingots were got, the oxygen contents are 12ppm and 25ppm. The silver plates were then rolled to thin sheets at room temperature. The reduction per pass is about 15%. Cold-rolled sheets were annealed at different temperatures and in different atmospheres.

Samples were analyzed using x-ray pole figures. (111), (200) and (220) incomplete pole figures were measured via the Shulz reflection method with Cu Kα radiation. The ODFs (orientation distribution function) were calculated using series expansion method of Bunge's system [11].

2. RESULTS

2.1 The influence of the oxygen content on the rolling and recrystallization textures

T.A.Gladstone and J.J.Wells suggested that low oxygen content in silver ingot promoted the formation of {110}<110> texture in Ag [8, 9]. But they did not report the quantitative oxygen content in the silver material. In this study, we have investigated the relationship between oxygen content and the texture in Ag quantitatively.

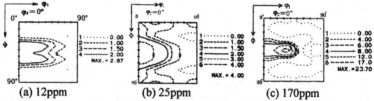

(a) 12ppm   (b) 25ppm   (c) 170ppm

Fig.1 The textures ($\varphi_2=0°$ ODF section) of Ag tapes with different oxygen contents after cold-rolling (total reduction is ~95%).

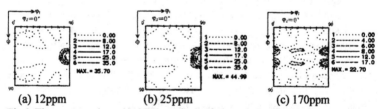

(a) 12ppm   (b) 25ppm   (c) 170ppm

Fig.2 The textures ($\varphi_2=0°$ ODF section) of Ag tapes with different oxygen contents after cold-rolling (total reduction is ~95%) and annealing at 900°C in air for 90 min.

Fig.1 and Fig.2 show the textures of Ag tapes with different oxygen contents (12, 25, and 170ppm) after cold-rolling (total reduction is 95% and the final thickness is 0.18mm) and annealing at 900°C in air for 90 min. respectively ($\varphi_2 = 0°$ ODF section). It was found that oxygen contained in Ag affects the cold rolling and recrystallization textures of Ag strongly. When the oxygen content is high, the rolling texture is very strong and has a {110}<211> ($\varphi_1 = 35°$, $\phi = 45°$ in $\varphi_2 = 0°$ section) main orientation and the annealing texture is ~{023}<110> ($\varphi_1 = 90°$, $\phi = 35°$, 55° in $\varphi_2 = 0°$ section) and ~{023}<100> ($\varphi_1 = 0°$, $\phi = 35°$, 55° in $\varphi_2 = 0°$ section). In contrast, for low-oxygen Ag, although under high deformation (95%), the rolling texture is weak and with large scatter, the strongest orientation concentrated in the ~{023}<uvw> ($\varphi_1 = 0~45°$, $\phi = 35°$, 55° in $\varphi_2 = 0°$ section) orientations (Fig.1a and b). After annealing in air, the principal orientation is {110}<110> ($\varphi_1 = 90°$, $\phi = 45°$ in $\varphi_2 = 0°$ section) orientation. (Fig.2a and b).

2.2 The influence of the total reduction on the rolling and recrystallization texture

Total reduction can always affect the texture of material strongly. In order to get the suitable reduction for the preparation of {110}<110> textured Ag substrate, 3.96 mm thick silver plate (oxygen content is 25 ppm.) was rolled to 0.51~0.18mm thin sheets. The total thickness reduction is 87%~95%. The samples were then annealed at 900°C in air for 90min.

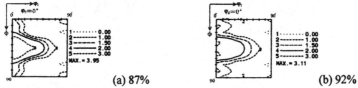

(a) 87%   (b) 92%

Fig.3 The $\varphi_2 = 0°$ ODF sections of the samples after cold-rolling under 87% and 92% reductions.

Fig.3 show the ODF sections measured from the samples after cold-rolling under 87% and 92% reductions. It can be seen that the deformation textures of silver after cold-rolling scattered around the {110}<100> direction ($\varphi_1 = 0°$, $\phi = 45°$ in $\varphi_2 = 0°$ ODF section). As the total reduction increased from 84% to 92%, the strongest orientations changed form the {110}<100> to the ~(023)<uvw> ($\varphi_1 = 0~45°$, $\phi = 35°$, 55° in $\varphi_2 = 0°$ ODF section) and (032)<uvw> ($\varphi_1 = 0~45°$, $\phi =$ in $\varphi_2 = 0°$ ODF section) orientations. Both of which are equivalent orientations in cubic crystal.

Fig.4 show the textures after cold-rolled under 87% and 92% reductions and annealed at 900°C for 30 min. in air ($\varphi_2 = 0°$ ODF section and (111) pole figure). It was found that as the total reduction increased, the spread of the recrystallization texture was continuously reduced and the strength of the main orientation {110}<110> was improved obviously after annealing. When the total reduction reached 92%, a nearly single component {110}<110> biaxial texture was formed after annealing and the samples after rolling under 93%, 94% and 95% reductions and annealing have the same result.

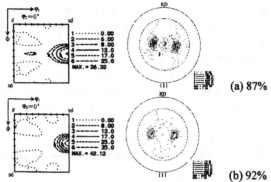

(a) 87%

(b) 92%

?ig.4 The $\varphi_2=0°$ ODF sections and (111) pole figures of the samples after cold-rolling under different total reductions and annealing in air at 900°C for 90 min.

2.3 The influence of the annealing temperature on the recrystallization textures

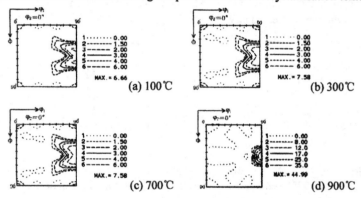

(a) 100°C

(b) 300°C

(c) 700°C

(d) 900°C

Fig.5 The textures after cold-rolling (total reduction is 95% and oxygen contain is 25ppm) and annealing at 100°C, 300°C, 700°C and 900°C for 90 min. in air (ODF $\varphi_2=0°$ section).

The cold-rolled samples (total reduction is 95%, the final thickness is 0.18mm and the oxygen content in it was 25 ppm.) were annealed at 100°C, 300°C, 500°C, 700°C and 900°C for 90 min. in air. The samples were heated slowly from room temperature to the annealing temperatures. This is an important measure to obtain a sharp expected texture. Fig.5 show the textures of some samples.

The results show that after annealed at low temperatures (100~300°C), the recrystallization textures of 95% cold rolled Ag tapes distribute near the $\{110\}<110>$ orientation. The strongest orientation is around $\{023\}<110>$ orientation ($\varphi_1 = 90°$, $\phi = 35°$, $55°$, in $\varphi_2 = 0°$ section). The strength of the $\{110\}<110>$ orientation increased continuously as the annealing temperature increased. When the annealing temperature reached $900°C$, a strong $\{110\}<110>$ biaxial texture with an in-plane FWHM of 8^0 was formed after annealing. This indicates that high annealing temperature is beneficial for the formation of $\{110\}<110>$ orientation.

2.4 The influence of annealing atmosphere on the recrystallization textures

Air, pure O_2, Ar and vacuum are used to study the influence of annealing atmosphere on the recrystallization textures. Cold-rolled samples (total reduction is 95%, the final thickness is 0.18mm and the oxygen content in it was 25 ppm) were annealed at $900°C$ for 90 min. in air, O_2, Ar and vacuum. The data shown in Fig.6 (a) and (b) were obtained from the samples annealed in Ar and vacuum respectively. The textures of Ag samples which annealed in air, O_2, Ar are nearly the same, which have the sharp $\{110\}<110>$ orientation. That means that annealing atmospheres such as air, O_2 and Ar have little influence on the recrystallization texture. But after annealing in vacuum, the recrystallization texture is quite different, there are two strong orientations which are $\{110\}<110>$ and $\{110\}<100>$ and weak $\{110\}<225>$ orientation ($\varphi_1 = 30°$, $\phi = 45°$ in $\varphi_2 = 0°$ section).

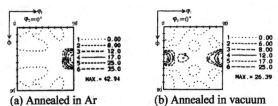

(a) Annealed in Ar   (b) Annealed in vacuum

Fig.6 The textures after cold-rolling (total reduction is 95% and oxygen contain is 25ppm) and annealing at $900°C$ for 90 min. in Ar (a) and vacuum (b) ($\varphi_2 = 0°$ ODF section).

3. DISCUSSION

It is well known that the annealing texture is strongly affected by cold-rolled texture. As the total reduction increases, the orientations in rolling texture concentrate around the $\{023\}<uvw>$, other orientations disappear gradually (Fig.3b). After annealing the orientation $\{110\}<110>$ become stronger with increasing reduction (Fig.4). When the reduction exceeds 92%, after annealing the texture shows nearly pure $\{110\}<110>$. This means that the orientation $\{110\}<110>$ may be transformed from the orientations around $\{023\}<uvw>$ and hence in order to obtain a pure $\{110\}<110>$ orientation, in cold rolling texture the orientations around $\{023\}<uvw>$ should be developed strongly, other orientations such as $\{015\}<001>$ ($\varphi_1=0°$, $\phi=10°$, in $\varphi_2=0°$ section) and $\{110\}<211>$ ($\varphi_1=35°$, $\phi=45°$, in $\varphi_2=0°$ section) can been seen as harmful ones.

Comparing the textures obtained at different annealing temperatures (Fig.5), it can be seen that when annealing temperature is lower, even at very low temperature e.g. 100℃, the recrystallization texture possesses always the strongest component near the $\{023\}<110>$ orientation and a weak orientation near $\{123\}<123>$ ($\varphi_1=85°$, $\phi=35°$ in $\varphi_2=25°$ section). This means that these orientations are the just completely recrystallized ones. After just recrystallization the $\{110\}<110>$ orientation in structure is weak compared to the other orientations. When further anneal at higher temperatures, e.g. higher than 700℃, $\{110\}<110>$ orientation becomes stronger. This result implies that the strong $\{110\}<110>$ orientation is caused by selected grain growth. Grain growth promotes the development of $\{110\}<110>$ orientation. Therefore the annealing temperature should be as high as possible e.g. 900℃ to ensure enough grain growth.

The influence of annealing atmosphere on the texture can be seen from Fig.6. Annealing in air, pure O_2 and Ar, a strong $\{110\}<110>$ texture can be obtained. From Fig.6b, it can be seen that the texture in Ag annealed in vacuum shows another kind one, there exist two strong texture components, $\{110\}<100>$ and $\{110\}<110>$. This effects indicates that under vacuum condition, the grain growth, in principle, promotes the development of the both orientations $\{110\}<100>$ and $\{110\}<110>$. This may be caused by the degassing effect occurred in grain boundaries of Ag thin sheet during vacuum annealing. Low gas content especially oxygen in Ag ingot before rolling is beneficial to the development of $\{110\}<110>$ oriented texture as is pointed out by some authors [8, 9], our experiment is agree with these results. Low oxygen content can affect only the rolling texture which has some difference compared with the texture obtained by rolling high oxygen content Ag. The rolling texture of low oxygen content Ag has a tendency to transform to $\{110\}<110>$ oriented texture during annealing. However low gas content can not restrain the formation of the orientation $\{110\}<100>$ and promote the selected grain growth to obtain a single $\{110\}<110>$ oriented texture during annealing as mentioned above. Therefore vacuum annealing is not a good way for obtaining single component $\{110\}<110>$ oriented texture. This means that gas e.g. O_2 is harmful for rolling texture but useful for annealing texture.

In our experiment the rolling texture of high oxygen content in Ag ingot (commercial Ag, ~170 ppm) shows a typical Brass type texture (Fig.1c), the strongest orientation concentrated around $\{110\}<211>$, after annealing in air the orientations changed to $\{023\}<110>$and $\{023\}<100>$ (Fig.2c). These orientations are not suitable for HTS deposition. When the oxygen content is reduced to 25 ppm, the orientations in rolling texture around $\{034\}<001>$ are well developed (Fig.1a,b), after annealing a strong $\{110\}<110>$ orientation is also developed (Fig.2a,b).

Fabrication of High Temperature Superconducto

4. CONCLUSION

In this study, we did a systematic investigation on the fabrication conditions of {110}<110> textured Ag foil. The obtained results indicate that, in order to obtain a suitable Ag substrate for HTS deposition, the oxygen content in silver ingot might not exceed 25 ppm, cold-rolling reduction should be higher than 92%, annealing temperature may be at about 900°C and the annealing treatment should be carried out in gaseous atmospheres such as inert gas. After such treatment one can obtain an Ag tape possessing strong {110}<110> texture.

ACKNOWLEDGEMENTS

This work is supported by of 863 project of China under Grant No.2002AA306221.

REFERENCES

[1] A.Goyal, D.P.Norton, J.D.Budai, M.Paranthaman, E.D.Specht, D.M.Kroeger, D.K.Christen, Q.He, B.Saffian, F.A.List, D.F.Lee, P.M.Martin, C.E.Kelabunde, E.Hartfield and V.K.Sikka, "High critical current density superconducting tapes by epitaxial deposition of $YBa_2Cu_3O_x$ thick films on biaxially textured metals," *Appl.Phys.Lett,* **69(12)** 1795-1797(1996).

[2] M.Yamazak, Y.Kudo, H.Yoshino, K.Ando and K.Ioue, "Preparation and Growth Mechanism of YBCO Thin film on Ag Substracte," pp.759-762, *8th International Symposium on Superconductivity (ISS'95),* Springer-Verlag, 1995.

[3] Y.Niiori, Y.Yamada, I.Hirabayashi, T.Fujiwara and K.Higashiyama, "In-plane aligned $YBa_2Cu_3O_{7-x}$ film on the MgO buffered Ag(100) substrate and {100}<001> cubic textured silver tape," *Physic C,* **301** 104-110 (1998).

[4] M.L. Zhou, H.S.Guo, D.M.Liu, T.Y.Zuo, L.H.Zhai, Y.L.Zhou, R.P.Wang, S.H.Pan and H.H.Wang, "Properties of $YBa_2Cu_3O_{7-\delta}$ films on textured Ag tapes," *Physic C,* **337** 101-105 (2000).

[5] D.M.Liu, M.L.Zhou, X.Wang, H.L.Suo, T.Y.Zuo, M.Schindle and R. Flükiger, "Epitaxial growth of biaxially oriented YBCO films on silver," *Supercond.Sci.Tech.* **14** 806-809 (2001).

[6] T.Yuasa, H.Kurosaki, S.Kim, T.Maeda, K.Higashiyama and I.Hirabayashir, "Fabrication of buffer layer for YBCO coated conductor on cube textured Ag substrate," *Physica C,* **357-360** 934-937 (2001).

[7] B.Ma, M.Li, R.E.Koritala, B.L.Fisher, S.E.Dorris, V.A.Maroni, D.J.Miller and U.Balachadran, "Direct deposition of YBCO on polished Ag substrates by pulsed laser deposition," *Physica C,* **377** 501-506 (2002).

[8] T.A.Gladstone, J.C.Moore, B.M.Henry, S.Speller, C.J.Salter, A.J.Wilkinson and C.R.M.Grovenor, "Control of texture in Ag and Ag-alloy substrates for superconducting tapes," *Supercond.Sci.Tech.* **13** 1399-1407 (2000)

[9] J.J.Wells, J.L.MacManus-Driscoll, J-Y Genoud, H.L.Suo, E Walker and R.Flükiger, "{110}<110> texture Ag ribbons for biaxially aligned $YBa_2Cu_3O_{7-x}$

coated conductor tapes," *Supercond.Sci.Tech.* **13** 1390-1398 (2000).

[10]T.Doi, M.Mori, H.Shimohigashi, Y.Hakuraku, K.Onabe, M.Okada, N.Kashima and Nagaya, "{110}<112> and {110}<110> textured Ag tapes for biaxially oriented YBa$_2$Cu$_3$O$_{7-x}$ coated conductors," *Physica C*, **378-381** 927-931 (2002).

[11] H.J.Bunge, "Texture Analysis in Material Science," Butterworths, London, 1982.

CHEMICALLY COATED BUFFER LAYERS DEPOSITED ON ROLLED NI SUBSTRATES FOR HTS COATED CONDUCTORS

Y.X. Zhou, S. Bhuiyan, H. Fang and K. Salama
Texas Center for Superconductivity and Advanced Materials
University of Houston, TX, 77204, USA

ABSTRACT

For fabricating long length $YBa_2Cu_3O_{7-x}$ (YBCO) coated conductors with high critical current density (J_c), easily scalable processing techniques for buffer layers are very important. In this paper, we report on the deposition of buffer layers using dip and spin coating on nickel substrates. The substrates were fabricated using the RABiTS approach. The solution was prepared from metal-organic precursors and was deposited on the Ni substrates using the dip and spin coating techniques. The films were annealed at 900°C—1150°C for 2 hours under 5% H_2-Ar gas flow. X-ray diffraction (XRD) of the buffer layers on the Ni tape shows a good out of plane alignment. The pole figure indicates a single cube-on-cube texture, and SEM observations reveal a continuous, dense and crack-free microstructure for the chemically coated buffers. These results offer the potential of further manufacturing coated conductors with high current density.

INTRODUCTION

Since the discovery of high temperature superconductors, enormous progress has been made in conductor development for the application of high temperature superconductivity in electric utilities. Two manufacturing techniques with the greatest potential for producing high-performance HTS wires at low cost, namely, the Oxide-Powder-in-Tube (OPIT)[1] and the coated-conductor methods[2,3,4,5] are being extensively studied. Both OPIT and coated conductor methods have one common feature, namely they can induce some form of texture in the final superconductor that allows for better connectivity between different grains in the microstructure. While the OPIT wires can be made in one-kilometer length, coated-conductor wires are much less mature and are being made in laboratories in lengths of a few meters. Coated-conductor techniques incorporate layers of material on flat tapes. Certain layers are textured, and this texture is then transformed to the HTS layer that is subsequently deposited. For these conductors, the substrate must be a flexible and smooth surface and the buffer layers have to be able to support oriented growth of YBCO, as well as have chemical and crystal compatibility with both the substrate and HTS layers. Therefore, the key technological challenge for coated conductors is how to fabricate a low-cost, scalable process to produce high quality substrate and buffer layers using a simple, easily scalable, and inexpensive method.

One of the most effective methods of fabricating coated conductors is the use of Rolling Assisted Biaxially Textured Substrate (RABiTS)[3] approach. In this work, the Ni substrates were textured using RABiTS approach and the buffer layers are deposited on rolled Ni substrates using chemical techniques from a solution-based precursor. XRD, SEM, and AFM were used to examine the quality of the textured substrates and buffer layers.

EXPERIMENTAL
Fabrication of Ni Substrates
The Ni strip was mechanically deformed by rolling to a degree greater than 95% total deformation using about 5% reduction per pass and reversing the rolling direction during each subsequent pass. It was annealed at about 1000°C for about 90 minutes to produce a sharp biaxial texture. Annealing was performed in flowing 5% H_2 in Ar. The procedure of rolling and annealing yielded the desired {100}<100> cubic textured Ni substrates[6].

Preparation of SrTiO₃ Precursor Solution
The solution preparation was carried out in air with strontium acetate (99.9%, Alfa Aesar) and titanium (IV) butoxide. Glacial acetic acid and acetyle acetone were used as chelating agents. 2-metoxyethanol and methanol were selected as solvents. Two concentrations of 0.25 M and 0.5 M were used for films deposited on the Ni substrate by spin coating and dip coating techniques respectively.

Preparation of CeO₂ Precursor Solution
The solution preparation was carried out in Ar atmosphere with Ce(IV)-metoxyethanol (99.9%, Alfa Aesar). Glacial acetic acid and acetyle acetone were used as chelating agents. 2-metoxyethanol and methanol were selected as solvents. Finally, the volume of the solution was adjusted with 2-methoxyethanol to 25ml to make a 0.1 M CeO₂ precursor solution.

Coating and Heating Procedure
A spin coating speed of 6000 rpm and a holding time of 60 seconds were used in the spin coating process. In the dip coating process, the textured Ni substrate was continuously pulled through the precursor solution at a constant rate of 100 cm/h.

The coated samples were annealed in a horizontal tube furnace at a temperature of 800°C---1200°C for 2 hours in flowing 5% H_2 in Ar. The flow chart and the typical heat treatment schedule of the MOD process of SrTiO₃

Fabrication of High Temperature Superconducto

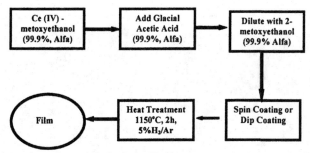

Fig. 1. Flow chart for MOD CeO$_2$ buffer layers on cube textured Ni substrate.

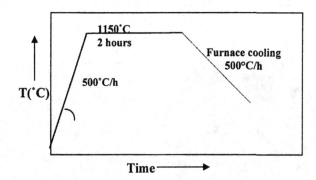

Fig. 2. Schematic illustration of the heat treatment schedule for CeO$_2$ buffer layers.

buffer layers were described in detail in reference[7]. The one of CeO$_2$ buffer layers on rolled Ni substrate are shown in figure 1 and figure 2.

Characterization Techniques

The samples were characterized in detail by X-ray diffraction (XRD). The out-of-plane alignment was measured by scanning the (200) planes of the samples. A Siemens General Area Detector Diffractions System (GADDS) with a pole figure goniometer was used to obtain pole figure measurements. Scanning Electron Microscopy (SEM) was used for imaging the substrates and the coated buffer surfaces.

RESULTS AND DISCUSSIONS

SrTiO$_3$ Buffer Layer

Fig. 3(a) shows the θ-2θ scans for SrTiO$_3$ buffer layers coated on rolled Ni substrates. As it is seen in the figure, there are two strong peaks at 2-theta angles of 46.5^O and 51.8^O, which correspond to the SrTiO$_3$ (200) peak and Ni (200) peak respectively. These two strong peaks reveal the presence of a sharp out-of-plane texture for both the rolled Ni substrate and the SrTiO$_3$ buffer layer. It is obvious

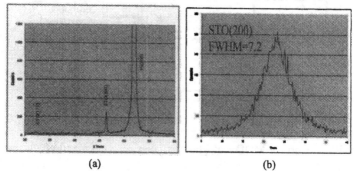

(a)   (b)

Fig. 3. XRD patterns of SrTiO₃ buffer layer coated on rolled Ni substrate.
(a) 2 Theta Scans (b) Rocking Curve.

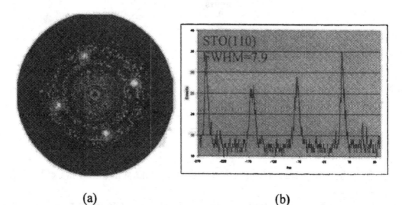

(a)   (b)

Fig. 4. (a) (110) x-ray pole figure of SrTiO₃ buffer layer shows a single cube-on-
cube epitaxy. (b) X-ray phi scans of SrTiO₃ buffer layer.

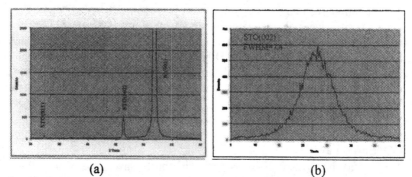

(a)   (b)

Fig. 5 XRD patterns of two coats of SrTiO₃ buffer layer deposited on rolled Ni
substrate.(a) 2 Theta Scans (b) Rocking Curve.

Fabrication of High Temperature Superconductor

that the intensities of the (110) and (111) peaks of SrTiO₃ buffer layer are negligible compared to the intensity of its (200) peak. This suggests that these buffer layers have a strong tendency towards preferred orientation, which was further confirmed using x-ray rocking curve analysis shown in Fig. 3(b). The full width at half-maximum (FWHM) values of out of plane alignment for SrTiO₃ (200) peaks are 7.2°. Figure 4(a) shows a (111) X-ray diffraction pole figure of the epitaxially textured SrTiO₃ buffer layer on rolled Ni substrates. As can be seen, four small poles located on the correct positions on the pole figure plot, which confirms that the SrTiO₃ phase grows epitaxially on rolled Ni substrate.

The film thickness of a single coat was determined to be ~50 nm. In order to increase the thickness, a thicker film was fabricated by repeating the coating and heating process. Fig. 5 shows the XRD scans for two coats SrTiO₃ buffer layer deposited on rolled Ni substrates. This figure also showed a strong peak at 46.5° from the SrTiO₃ (200) reflection and a very weak peak at 2 theta of 32.4° from the SrTiO₃ (110) reflection, which indicated that two coats SrTiO₃ buffer layer still had a good out-of-plane alignment on rolled Ni substrate. The rocking curve is shown in Fig. 5(b) and the FWHM is 7.4°.

To compare the spin coating and dip coating techniques, we also characterized the SrTiO₃ buffer layers deposited on rolled Ni substrate using dip coating technique. The layers were characterized by the θ-2θ scans and pole figure analysis, plots of which are not shown because they are similar to those for the buffer layer deposited on rolled Ni substrate using spin coating. This is confirmed by comparing the in-plane and out-of-plane alignments of the both buffer layers.

CeO₂ Buffer Layer

Among the buffer layers, CeO₂ is usually considered as another high quality material, because of the good lattice match of the (100) plane with the YBCO (110) plane, the comparable thermal expansion coefficient ($11.6*10^6$ K⁻¹ for CeO₂, $12-13*10^6$ K⁻¹ for YBCO), the oxidation protection metallic substrate and the chemical compatibility with YBCO and many other material substrates[8]. So far, there are several methods which are being explored for the fabrication of CeO₂ buffer such as PLD, MOCVD, and Sputtering [4]. Compared to these techniques, MOD[9,10,11] offers several attractive features which have been discussed before. A typical θ-2θ scans for a CeO₂ buffer layer deposited on a textured Ni (100) substrate is shown in Fig. 6 (a). The strong CeO₂ (200) signal revealed the presence of a good out-of-plane texture. Fig. 6(b) shows the ω scans for as grown CeO₂ buffer on a textured Ni substrate. The full width at half maximum value for CeO₂ (200) is 7.8°. From these XRD results, it can be concluded that CeO2 can be grown on a rolled Ni substrate with a good out-of-plane texture.

SEM Micrographs

SEM micrographs for SrTiO₃ buffer layers on rolled Ni substrate are discussed in reference[7], which indicates that the SrTiO₃ buffer layer on textured Ni substrate has very fine grains around 150 nm and it is continuous as well as crack free. Fig.

7 shows an SEM micrograph for the CeO$_2$ buffer layer coated on rolled Ni substrate. As can be seen from the SEM micrograph, the CeO2 buffer layer is dense, crack free and continuous. Efforts are underway to grow YBCO directly on these SrTiO$_3$ and CeO$_2$ buffered Ni substrates.

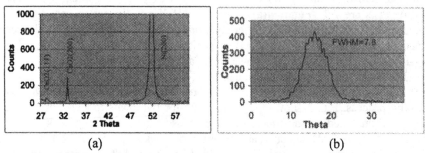

(a)   (b)

Fig. 6 XRD patterns of CeO$_2$ buffer layer coated on rolled Ni substrate.
(a) 2 Theta Scans (b) Rocking Curve.

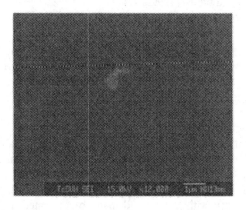

Fig. 7. SEM micrograph for the CeO$_2$ buffer layer coated on rolled Ni substrate.

CONCLUSIONS

Chemically processed SrTiO$_3$ and CeO$_2$ buffer layers were produced from metal-organic precursor based solutions. The ω scans show both the SrTiO$_3$ and CeO$_2$ buffer layers have a good out-of-plane alignment on rolled Ni substrate. The pole figures indicate a single cube-on-cube texture, and, SEM observations reveal a continuous, dense and crack-free microstructure. Efforts are underway to grow YBCO directly on SrTiO$_3$ and CeO$_2$ buffered Ni substrates.

ACKNOWLEDGMENT

This work was financially supported by the Air Force Office of Scientific Research as part of project F49620-01-1-0391, and the Texas Center for Superconductivity and Advanced Materials.

REFERENCES

[1]U. Balachandran, A. N. Iyer, P. Haldar, and L. R. Motowidlo, JOM 45 (1993) 54.

[2]Norton, D. P. et al., Science 274 (1996) 755.

[3]A. Goyal, D.P. Norton, J.D. Budai, M. Paranthaman, E.D. Specht et.al. Appl. Phys. Lett. 69 (1996) 1795.

[4]Iijima Y, Onabe K, Futaki N and Ikeno Y, Appl. Phys.Lett. 60 (1992) 769.

[5]K. Hasegawa, N. Yoshida, K. Fujino, H. Mukai, K. Hayashi, K. Sato, Y. Sato, S. Honjo, T. Ohkuma, et.al Advaces in Superconductivity IX, Proceedings of 9th International Sysposium on Superconductivity (ISS'96), Sappora, Japan, 1996, p.745.

[6]Y.X.Zhou, T. Rizwan and K. Salama, IEEE transactions on applied superconductivity, 2003(in press).

[7]Y.X.Zhou, S. Bhuiyan, S.Scruggs, H. Fang, M. Mironova and K.Salama, Supercond.Sci.Tech. 16 (2003) 1.

[8]I. Van Driessche, G. Penneman, C. De Meyer, I. Stambolova, E. Bruneel, S. Hoste, Engineering Materials, V206-213(2002) 479.

[9]M Paranthanam, D F Lee, A Goyal, E D Specht, P M Martin, X Cui, J E Mathis, R Feenstra, D K Christen and D M Kroger, Supercond. Sci. Technol. 12 (1999) 319-325.

[10]Srivatsan Sathyamurthy and Kamel Salama, Supercond. Sci. Technol. 13(2000) L1-L3.

[11]H. Okuyucu, E. Celik, M. K. Ramazanoglu, Y. Akin, I. H. Mutlu, W. Sigmund, J. E. Crow, Y. S. Hascicek, IEEE Trans. Appl. Supercond. 11 (2001) 2889.

Fabrication of High Temperature Superconducto

DEVELOPMENT OF CONDUCTIVE $La_{0.7}Sr_{0.3}MnO_3$ BUFFER LAYERS FOR Cu-BASED RABiTS

Tolga Aytug, M. Parans Paranthaman, Amit Goyal, Albert Gapud, Noel Rutter, Hong Ying Zhai and David K. Christen
Oak Ridge National Laboratory
1 Bethel Valley Road
Oak Ridge, TN 37831

ABSTRACT

For the development of $YBa_2Cu_3O_{7-\delta}$ (YBCO)-based coated conductors for electric power applications, it is desirable to electrically and thermally stabilize the high temperature superconducting (HTS) coating during its operation. In addition, use of non-magnetic substrates is an equally important consideration in conductor configurations, in order to reduce ferromagnetic hysteresis energy loss in ac applications. We have developed a conductive buffer layer of $La_{0.7}Sr_{0.3}MnO_3$ (LSMO) on biaxially textured non-magnetic Cu substrates to electrically couple the HTS layer to the underlying metal substrate. Property characterizations of YBCO films on LSMO/Ni/Cu multilayer architecture revealed good electrical connectivity over the entire structure.

INTRODUCTION

Potential applications of HTS coated-conductors involve areas of electric utility and power industry. For these applications, coated conductor tapes must be stabilized against thermal runaway in the event of an over-current situation. The ideal solution to avoid thermal runaway is to electrically couple the thin HTS layer to the thick underlying metal substrate through a conductive buffer layer, which reduces the overall resistivity in the event of a transient loss of superconductivity. As long as the power generated per unit surface area of conductor, due to Joule heating, remains sufficiently low, (eg. relative to the critical heat flux of boiling cryogen, or the thermal diffusivity of the system [1,2]) then event can be controlled and superconductivity should eventually be restored.

To date, much of the work reported using the rolling assisted biaxially textured substrates (RABiTS) [3] technique has utilized high purity Ni (99.99%) as the texture material. However, the ferromagnetism (FM) of pure Ni presents significant challenges in long length manufacturing where AC losses are an issue. On the other hand, Cu is a lower-cost, lower-resistivity and non-magnetic alternative substrate material for the production of long length RABiTS-based coated conductor tapes.

In this study, we assess the viability of $La_{0.7}Sr_{0.3}MnO_3$ (LSMO) as a *conductive* buffer-layer on biaxially Cu tapes. The structural, electrical and superconducting properties of these short prototype conductors are reported.

EXPERIMENTAL

Biaxially textured Cu substrates were obtained from randomly oriented high purity (99.99%) Cu bars, which are first mechanically deformed by cold-rolling and then annealed in vacuum at 800 °C for 1 hour, to obtain the desired (100) cube texture. For some samples, a thin ~ 2μm layer of Ni was first deposited. Deposition of LSMO films was conducted by *rf*-magnetron sputtering in a mixture of reducing forming gas (96%Ar + 4%H$_2$) and $2x10^{-5}$ Torr of H$_2$O. The substrate temperature was kept around 550-625 °C. The YBCO films were grown by pulsed laser deposition (PLD), using a KrF excimer laser system operated at an energy density of ≈2 J/cm^2 at 780 °C in 120 mTorr of O$_2$. Typical thicknesses of the YBCO films were 200 nm. The crystallographic orientation and texture of the films were characterized by X-Ray Diffraction (XRD), and surface microstructural investigations were conducted on a JOEL model, JSM-840 scanning electron microscope (SEM). A standard four-probe technique was used to evaluate the electrical properties, including T$_c$, and J$_c$ of the YBCO films.

RESULTS

Typical θ-2θ XRD spectra for a 600 nm thick LSMO layer deposited directly on Cu substrate and the subsequent YBCO coating are shown in Figs. 1a and 1b, respectively. The patterns indicate only (00*l*) reflections from both the YBCO and the LSMO layers, demonstrating that the YBCO/LSMO/Cu are *c*-axis-oriented. There is no evidence of CuO/Cu$_2$O formation at the interface after buffer layer deposition. However, after YBCO deposition, additional peaks at around 37° and 43° appeared, indicating the formation of a large amount of copper oxide in the sample. To investigate the details of this issue, SEM analysis has been made. Figure 2a illustrates the surface morphology of the same YBCO as in Fig. 1b. Figure 2b is the higher magnification image of the boxed region. It can be clearly seen from these micrographs that the YBCO surface is covered with cracks, lying parallel to the sample, and through these cracks Cu-O diffuses in to the sample. After some detailed annealing studies, we deduce that the reason for the existence of cracks can be explained by the volume expansion (≈ 58 %) per Cu atom in Cu$_2$O, relative to that of the Cu in pure Cu matrix, and not because of the thermal expansion difference between the metal substrate (α = 18x10^{-6}/ °C) and the oxide layers (α ≈11x10^{-6}/ °C). During the YBCO deposition, oxygen diffuses through the LSMO into the Cu substrate resulting in large amount of interfacial Cu-O

Fabrication of High Temperature Superconducto

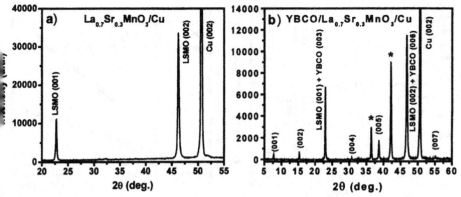

Fig. 1 Typical XRD θ-2θ scans (a) for LSMO/Ni. (b) for 200 nm YBCO film on the same LSMO/Cu multilayer. (* indicates Cu_2O impurities)

formation with associated volume expansion, leading to stress build-up on the LSMO layer, and eventually to crack development. In order to overcome this issue and to achieve an electrically conductive architecture we have chosen the

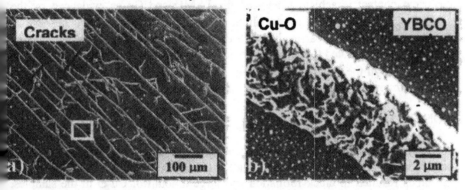

Fig. 2 SEM micrograph of (a) the surface morphology of the YBCO film grown on the LSMO/Cu substrate, and (b) the magnified image of the boxed region in part a.

alternative route using protective Ni overlayers on Cu substrates. Previously we have demonstrated high quality growth of La-based manganese oxide layers on Ni-based templates [4]. Figure 3a shows the θ-2θ XRD pattern for the LSMO/Ni/Cu, while Fig. 3b plots the pattern for the YBCO film on the same LSMO/Ni/Cu architecture. The presence of only (00l) reflections indicates that all layers are oriented with its c-axis normal to the plane of the substrate. After YBCO deposition, presence of a small amount of NiO is detected, the consequence of which is explained in Fig. 4. Note that the T_c of the sample is 90.4 K and the 77 K self-field J_c value is 2.3×10^6 A/cm^2. In this figure, we compared the net resistivity (ρ_{net}) versus temperature behavior of YBCO/LSMO/Ni/Cu

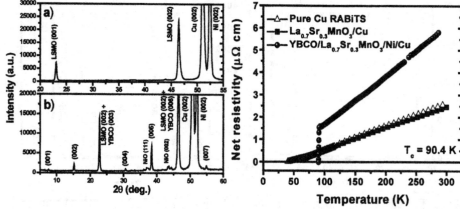

Fig. 3 XRD θ-2θ patterns (a) for LSMO/Ni/Cu (b) for a YBCO film on LSMO/Ni/Cu multilayers.

Fig. 4 shows the temperature dependent resistivities of pure Cu LSMO/Cu, and YBCO/LSMO/Ni/Cu.

sample with data for the LSMO/Cu as well as with the pure textured Cu. Clearly, electrical contact is achieved between the LSMO and Cu as evidenced by the similar ρ_{net}-T behaviors. However slight increase in ρ_{net}, after YBCO deposition, indicates some contact barrier between the oxide layers and the metal substrate, which can be attributed to the presence of NiO at the substrate-buffer interface, as was detected in the XRD spectra of Fig. 3b. Nevertheless, ρ_{net} values < 2 $\mu\Omega$-cm at LN_2 temperatures should provide significant stability to the entire structure.

CONCLUSION

In summary, we have demonstrated the successful fabrication of a conductive multilayer architecture on biaxially textured Cu substrates having the layer sequence of YBCO/LSMO/Ni/Cu. Since the YBCO growth on LSMO buffered Cu tapes resulted in crack formation and large amount of Cu-O in the samples, additional Ni overlayer is used to provide a barrier for Cu oxidation and/or diffusion. Electrical property characterization revealed good electrical coupling between the YBCO coating and the LSMO/Ni/Cu.

REFERENCES

1. C. Cantoni, T. Aytug, D.T. Verebelyi, M. Paranthaman, E.D. Specht, D.P. Norton, and D.K. Christen, IEEE Trans. on Appl. Supercond. 11, 3309 (2001).
2. Y. Fu, O. Tsuamoto, and M. Furuse, IEEE Trans. on Appl. Supercond. (2003).
3. A. Goyal, D.P. Norton, J.D. Budai, M. Paranthaman, E.D. Specht, D.M. Kroeger, D.K. Christen, Q. He, B. Saffian, F.A. List, D.F. Lee, P.M. Martin, C.E. Klabunde, E. Hatfield, and V.K. Sikka, Appl. Phys. Lett. 69, 1795 (1996).
4. T. Aytug, M. Paranthaman, H.Y. Zhai, H.M. Christen, S. Sathyamurthy, D.K Christen, and R.E. Ericson, J. Mater. Res. 17, 2194 (2002).

PULSED LASER DEPOSITION OF YBCO WITH YTTRIUM OXIDE BUFFER LAYERS

Rama M. Nekkanti,* Paul N. Barnes, Lyle B. Brunke,* Timothy J. Haugan, Nick A. Yust, Iman Maartense,** John P. Murphy,** Srinivas Sathiraju,[†] Juliana M. Evans, and Justin C. Tolliver
Air Force Research Laboratory
AFRL/PRPG, Building 450
Wright-Patterson AFB, OH 45433
*UES, Inc., **UDRI, [†]NRC Fellow

Kenneth R. Marken, Jr.
Oxford Superconducting Technology
600 Milik Street
Carteret, NJ 07008

ABSTRACT

Textured metallic substrate based HTS coated conductors with the YBCO/CeO$_2$/YSZ/CeO$_2$/Ni architecture have already been shown to exhibit high current densities. The CeO$_2$ seed layer can effectively minimize the formation of NiO during the initial deposition on Ni and the CeO$_2$ cap layer provides good lattice matching to the subsequent YBCO layer. However, there are reports of cracks developing in the CeO$_2$ seed layer after a thicker growth due to a lattice mismatch with Ni, which can lead to poor performance of the YBCO conductor. The present work explores an alternate approach by using yttrium oxide not only as the seed layer but also as the cap layer in place of CeO$_2$. In the literature, yttrium oxide films grown on nickel by electron beam evaporation processes were found to be dense and crack-free with good epitaxy. This is likely the first report of using Y$_2$O$_3$ as a seed as well as a cap layer within the YBCO coated conductor architecture on specimens being fabricated in a single chamber. Pulsed laser deposition was used to perform deposition of all layers. Preliminary experiments resulted in specimens with current densities of more than 1 MA/cm^2 at 77K in self field. Characterization of samples was accomplished using x-ray diffraction, both resistive and ac susceptibility derived T$_c$, and J$_c$ transport measurements.

INTRODUCTION

Coated high temperature superconductors based on textured substrate technology are in the forefront of developmental efforts towards long length manufacturing of coated conductor tapes for practical applications.[1-3] Significant effort is focused on the development of new metallic substrates based on nickel alloys which are strong, non-magnetic, and are well textured.[4-6] At the same time, different combinations of buffer layers are also being developed to transfer the almost perfect cube texture in the metallic substrate to the $YBa_2Cu_3O_{7-x}$ (YBCO) layer epitaxially. Improvement of the in-plane alignment in the YBCO layer through a suitable seed layer on metallic substrates is an impotant aspect of the YBCO coated conductor development. The main considerations for the buffer layer are the lattice matching, thermal expansion coefficient matching, chemical compatibility, and ease of deposition.

Previously, a lot of developmental work has concentrated on the YBCO/ CeO_2/YSZ/CeO_2/Ni architecture which resulted in current densities exceeding 1 MA/cm^2 on samples.[7] The initial seed layer of CeO_2 has been very effective in preventing the formation of NiO, which might otherwise affect the epitaxy of subsequently applied layers. YSZ is a good diffusion barrier preventing the diffusion of nickel into the superconductor. A cap layer is deposited on YSZ to suppress the growth of $BaZrO_3$ at the interface between YSZ and YBCO. CeO_2 is a favored cap layer for YSZ due to its minimal mismatch with both YSZ and YBCO. Even so, interaction between the CeO_2 and YBCO layer may sometimes occur resulting in the formation of $BaCeO_3$ which can lead to a lower current density in the YBCO.[8] Also, the CeO_2 seed layer cracks after thicker growth due to the lattice mismatch with nickel.

Yttrium oxide (Y_2O_3) has a better lattice match with nickel and can reduce cracking problems. As such, yttrium oxide has successfully demonstrated replacement for the CeO_2 as the seed layer, alleviating the potential cracking problems even though CeO_2 is still preferred for the cap layer.[3] In the literature, there are successful reports of deposition of thick biaxially textured yttrium oxide layers on textured nickel by electron beam evaporation.[9] Ichinose et al. have described the process-related crystalline alignment and microstructure of Y_2O_3 buffer layers deposited under various deposition conditions by e-beam evaporation.[10] Paranthaman et al have deposited other buffer layers by sputtering on top of the yttrium oxide followed by pulsed laser deposition of the YBCO layer which resulted in high current densities (J_c ~1.8 x 10^6 A/cm^2).[11] The present work deals with yttrium oxide replacing CeO_2 as the seed as well as the cap layer. All layers, including YSZ and YBCO, are deposited in-situ in the same chamber by pulsed laser deposition.

EXPERIMENTAL

Textured nickel substrates were obtained from Oxford Instruments and the processing details have been presented elsewhere.[6] The substrates have an in-plane alignment of 7.2° FWHM and out-of-plane alignment of 8.4°. The various

Fabrication of High Temperature Superconducto

oxide buffer layers and YBCO were deposited using pulsed laser deposition system in a Neocera chamber with a Lambda Physik (Model LPX 305i) excimer laser operating at the KrF wavelength of 248 nm. Specimens were fabricated to different stages in the YBCO/CeO$_2$/YSZ/CeO$_2$/Ni and YBCO/Y$_2$O$_3$/YSZ /Y$_2$O$_3$/Ni architecture to study the texture, smoothness and microstructure of the deposited layers. The background pressure in the chamber was brought down to < 10^{-6} torr pressure after mounting the specimens on the 2" diameter substrate heater. Specimens were then heated from room temperature to 750 °C in 180 mtorr atmosphere of Ar+4% H$_2$ (forming gas) gas mixture to prevent oxidation of the nickel substrate.

The layers were applied in-situ in the following manner. After a soaking period of 10 min, the Y$_2$O$_3$ (or CeO$_2$) seed layer was deposited at a laser energy of 625 mJ in the Ar+H$_2$ gas mixture for 3 min at a 4 Hz laser repetition rate. The laser spot size on target was 4.6 mm^2. The gas fill in the chamber was then evacuated and the chamber was pumped down to a pressure of 10^{-6} torr; the deposition of Y$_2$O$_3$ (or CeO$_2$) was continued for an additional 1.5 min. Oxygen gas was then introduced in to the chamber and after stabilizing the pressure at 10^{-4} torr, the Y$_2$O$_3$ (or CeO$_2$) layer was further deposited for 2 min. The temperature was then increased from 750 °C to 780 °C, and the YSZ buffer layer was deposited for 20 min in the oxygen atmosphere at a laser energy of 650 mJ using a 10 Hz frequency. A cap layer of Y$_2$O$_3$ was then deposited at a laser energy of 625 mJ and a 4 Hz repetition rate for 2 min. The oxygen pressure was subsequently increased to 600 mtorr and the superconducting YBCO layer was then deposited on the buffer layers.

The as deposited films were analyzed by detailed x-ray diffraction studies. Two theta scans were accomplished by using a Rigaku x-ray diffractometer. A Philips MRD with four circle diffractometry was used to study the crystalline alignment of the substrate, buffer layers, and the superconductor by means of omega, phi and psi scans. The microstructure of the various films was evaluated under scanning electron microscopy (SEM) to study the surface roughness and morphology of the film structure. Atomic force microscopy (AFM) was used to characterize the roughness of the buffer layers. The quality of YBCO was evaluated by x-ray scans as well as by ac susceptibility measurements to determine the critical transition temperatures (T$_c$'s). Electrical property characterizations were made using a standard four-probe technique with a 1 μV/cm criterion to determine the critical current (I$_c$).

RESULTS AND DISCUSSION

The I-V plot of the current measured in the specimen RN-68 at liquid nitrogen temperature, representative of several good samples using Y$_2$O$_3$ as the seed and cap layer, indicated the specimen carried a critical current of 18 A which is equivalent to 1.2 MA/cm^2 of critical current density. Specimen RN-36 at 77K, representative of several good samples using CeO$_2$ as the seed and cap layer, carried an I$_c$ of 15 A equivalent to a J$_c$ of 1.0 MA/cm^2. Figure 1 shows ac susceptibility data obtained for RN36 but is representative of both architectures indicating high T_c. The ac susceptibility plots of the YBCO specimen shown in Figure 1 exhibit an onset T_c of 90.1 K along with lower transition widths and peak shifts of the χ'' for varying magnetic field indicating strong inter-grain coupling and a potentially higher J_c.[12]

Thicker ceria seed layers induced cracking although yttria layers did not crack even as thickness increased. Figure 2 provides an image of the surface of the YBCO microstructure by SEM and AFM. The underlying grain boundaries of the Ni substrate are transferred to the final YBCO layer.

The x-ray theta-two theta scans on YBCO/Y_2O_3/YSZ/Y_2O_3/Ni and YBCO/CeO_2/YSZ/CeO_2/Ni as well as the intermediate buffer layers, showed sharp (00l) peaks indicating excellent c-axis texture in buffer layers which was carried over to the superconducting YBCO layer. Phi scans of the different layers on nickel indicate excellent in-plane alignment of the various layers although slightly better for the ceria architecture: YSZ = 8.3°, Y_2O_3 = 8.0°, and YBCO = 10° for the yttria architecture and for the ceria architecture CeO_2 = 6.7°, YSZ = 6.7°, YBCO = 7.1°. The FWHM of the buffer layers and YBCO did not change much from layer to layer indicating good epitaxy. Refer to Figures 3 and 4 for Psi scans of the two specimens. Figures 3 and 4 provide a comparison between the YSZ and YBCO layer of the samples.

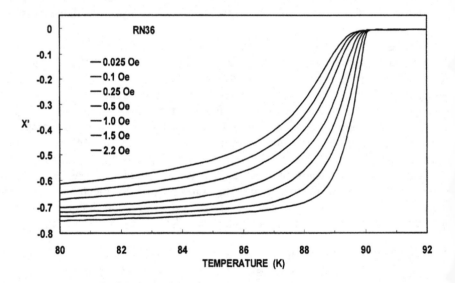

Figure 1. AC susceptibility data for the YBCO/CeO_2/YSZ/CeO_2/Ni architecture on a given sample. The different curves result from the different applied fields listed in the legend—the field increases from right to left.

Fabrication of High Temperature Superconducto

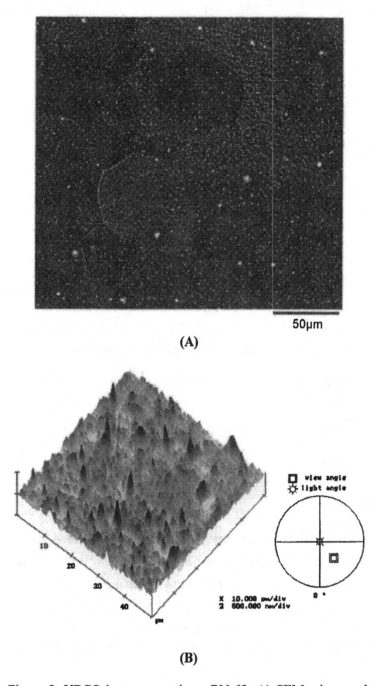

(A)

(B)

Figure 2. YBCO layer on specimen RN-68: A) SEM micrograph displaying a large area of the surface microstructure, B) AFM picture showing the surface morphology of a 50 μm x 50 μm area .

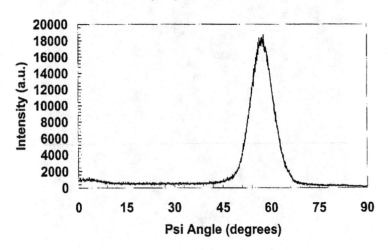

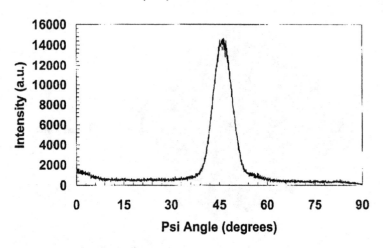

Figure 3. Psi Scan on (111) Peak of YSZ and YBCO layer of specimen RN-36 for the YBCO/CeO$_2$/YSZ/CeO$_2$/Ni architecture.

Fabrication of High Temperature Superconductor

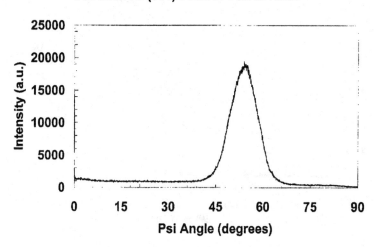

Psi Scan on (111) Peak of YSZ of RN68

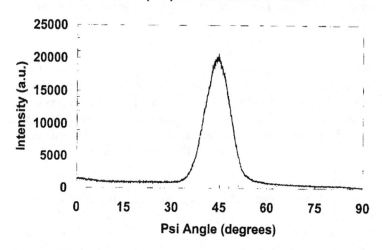

Psi Scan on (111) Peak of YBCO of RN68

Figure 4. Psi Scan on (111) Peak of YSZ and YBCO layer of specimen RN-68 for the YBCO/Y_2O_3/YSZ/Y_2O_3/Ni architecture.

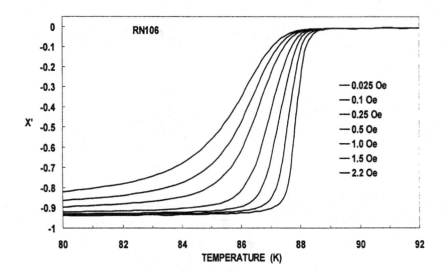

Figure 5. AC susceptibility data for the YBCO/Y$_2$O$_3$/Ni architecture for a given sample. The different curves result from the different applied fields listed in the legend—the field increases from right to left.

Additionally, a single buffer layer of Y$_2$O$_3$ applied to the Ni substrate was also experimented with. The ac susceptibility plots of the YBCO/Y$_2$O$_3$/Ni specimen showed a reasonable T$_c$ (89 K onset) along with slightly broader transition widths and peak shifts of the χ'' for varying magnetic fields compared to those of the traditional YBCO coated conductor architecture. More work needs to be done to optimize the performance using a single Y$_2$O$_3$ buffer layer.

CONCLUSION
Yttrium oxide was successfully incorporated as both a seed and a cap layer in fabricating the YBCO coated conductor specimens on a textured nickel substrate in a single chamber using pulsed laser deposition. Good epitaxy was observed in all the deposited layers of the YBCO/Y$_2$O$_3$/ YSZ/Y$_2$O$_3$/Ni architecture leading to a high T$_c$ (~91 K) and self-field J$_c$'s of more than 1 MA/cm^2 on multiple samples. The microstructure of the yttrium oxide and the superconducting YBCO layers were dense, crack-free, and continuous with uniform coverage across the sample.

REFERENCES
[1]A.C. King, S.S. Shoup, M.K. White, S.L. Krebs, D.M. Mattox, T. Polley, N. Darnell, K.R. Marken, "The progress made using the combustion chemical vapor deposition (CCVD) technique to fabricate YBa$_2$Cu$_3$O$_{7-x}$ coated

Fabrication of High Temperature Superconducto

conductors," IEEE Trans. on Appl. Superconductivity, **13** [2], pp. 2643 – 2645 (2003).

[2] R.A. Hawsey and A.W. Murphy, *ORNL Superconducting Technology Program for Electric Power Systems, Annual Report for FY 2002,* ORNL/HTSPC-14 (2003).

[3] D.T. Verebelyi, U. Schoop, C. Thieme, X. Li, W. Zhang, T. Kodenkandath, A.P. Malozemoff, N. Nguyen, E. Siegal, D. Buczek, J. Lynch, J. Scudiere, M. Rupich, A. Goyal, E.D. Specht, P. Martin, and M. Paranthaman, "Uniform performance of continuously processed MOD-YBCO-coated conductors using a textured Ni–W substrate," Superconductivity Science & Technology, **16** pp. L19-L22 (2003).

[4] R. Nekkanti, V. Seetharaman, L. Brunke, I Maartense, D. Dempsey, G. Kozlowski, D. Tomich, R. Biggers, T. Peterson, and P. Barnes, "Development of Nickel Alloy Substrates for YBCO Coated Conductor Applications," IEEE Trans. on Appl. Superconductivity, **11**, pp. 3321-3324 (2001).

[5] M. Paranthaman, F. A. List, S. Sathyamurthy, T. Aytug, S. Cook, A. Goyal, and D. M. Kroeger, "Strengthened Ni-W Substrates Produced in 35-m Lengths with Smooth Surfaces and Good Texture," pp.1-11 to 1-12 in *ORNL Superconducting Technology Program for Electric Power Systems, Annual Report for FY 2002,* Edited by W.S. Koncinski, ORNL Technical Report HTSPC-14, 2003.

[6] K. Marken, B. Czabaj, S. Hong, S. Shoup, M. White, S. Krebs, N. Darnell, T. Polley, D. Mattox, D. Bryson, F. Fortunato, M. Carlin, C. Koehly, and A. Hunt, "Coated Conductor R&D using RABiTS and Combustion Chemical Vapor Deposition," Dept. of Energy Superconductivity for Electric Systems Annual Peer Review, Washington, DC, July 24, 2003.

[7] C.Y. Yang, S.E. Babcock, A. Goyal, M. Paranthaman, F.A. List, D.P.Norton, D.M. Kroger, A. Ichinose, "Microstructure of electron-beam-evaporated epitaxial yttria-stabilized zirconia/CeO_2 bilayers on biaxially textured Ni tape," Physcia C, **307**, pp. 87-98 (1998).

[8] T.G. Holesinger, S.R. Foltyn, P.N. Arendt, Q. Jia, P.C. Dowden, R.F DePaula, and J.R. Groves, "A Comparison of Buffer Layer Architectures on Continuously Processed YBCO Coated Conductors Based on the IBAD YSZ Process," IEEE Trans. on Appl. Superconductivity, **11** [1], pp. 3359 – 3364 (2001).

[9] M. Paranthaman, D.F. Lee, A. Goyal, E.D. Specht, P.M. Martin, X. Cui, J.E. Mathis, R. Feenstra, D.K. Christen, and D.M. Kroeger, "Growth of biaxially textured RE_2O_3 buffer layers on rolled-Ni substrates using reactive evaporation for HTS-coated conductors," Supercond. Sci. Technol., **12**, pp. 319–325 (1999).

[10] A. Ichinose, C. Yang, D. Larbalestier, S. Babcock, A. Kikuchi, K. Tachikawa, S. Akita, " Growth Conditions and microstructure of Y_2O_3 buffer layers on Cube Textured Ni", Physica C, **324**, pp.113-122, (1999).

[11] D.F. Lee, M. Paranthaman, J.E. Mathis, A. Goyal, D.M. Kroeger, E.D. Specht, R.K. Williams, F.A. List, P.M. Martin, C. Park, D.P. Norton, D.K.

Christen, "Alternative Buffer Architectures for High Critical Current Density YBCO Superconducting Deposits on Rolling Assisted Biaxially-Textured Substrates," Jpn. J. Appl. Phys., **38**, pp. L178-L180 (1999).

[12]I. Maartense and A. K. Sarkar, "Annealing of pressure-induced structural damage in superconducting Bi-Pb-Sr-Ca-Cu-O ceramic," J. Mater. Res., **8**, 2177 (1993).

Fabrication of High Temperature Superconducto

BSCCO–Based Conductors, MgB$_2$ and Other HTS Materials

HIGH TRANSPORT PROPERTIES IN IRON-CLAD MGB$_2$ WIRES AND TAPES

H. Fang, S. Padmanabhan, Y.X. Zhou, P.T. Putman, and K. Salama
Texas Center for Superconductivity and Advanced Materials (TcSAM) and
Department of Mechanical Engineering,
University of Houston, Houston, TX 77204-4006

ABSTRACT

Iron-clad MgB$_2$ superconducting wires and tapes were fabricated using the standard powder-in-tube (PIT) method. The starting precursor was ultra-fine Mg and B mixture powder prepared by high-energy ball milling. Very good grain connections as well as grain refinement were obtained. Under 1.5 Tesla external magnetic field, J_c of MgB$_2$ tape is 1.07×10^5 A/cm^2 and 6.54×10^3 A/cm^2 at 20 K and 30 K, respectively. An extrapolation to zero field at 20 K gives a J_c of 3.0×10^5 A/cm^2. A solenoid with a diameter of 12 mm and 12 turns made from MgB$_2$ wires demonstrated the potential for large-scale applications. Under 1 Tesla, the critical current of this solenoid is 159 A and 110 A at 20 K and 30 K, respectively.

INTRODUCTION

The discovery of superconductivity at 39 K in basic intermetallic magnesium diboride by Nagamatsu et al. [1] triggered a large volume of research on the basic physics of this material as well as its applied superconductivity. The absence of weak-links between micro-crystals or grains of this material reported by Larbalestier et al. [2] demonstrated its advantage over high T_c oxide superconductors. Despite its relative low T_c compared to its HTS peers, weak-link free boundaries and low manufacturing cost make MgB$_2$ a promising candidate for power applications such as fault-current controllers, large motors, generators, SMES and transformers [3]. In addition, progress in cryocooler development and the accessibility of cryogen free cooling in the temperature range of 20 ~ 30 K may actually promote the application of MgB$_2$. These applications require the development of MgB$_2$ tapes and wires with superior transport currents under external magnetic field. Currently the powder-in-tube method is widely used to fabricate MgB$_2$ tapes and wires [4]-[10], where MgB$_2$ powder or a mixture of Mg and B powders with the stoichiometric composition is packed into a metal tube. The powder-filled metal tube is then cold worked (swaging, drawing and rolling) into tapes or wires. A heat treatment at 600 ~ 1000 °C is usually applied to the cold-worked tapes or wires to obtain superconductivity. The sheath material's toughness and chemical compatibility with MgB_2 are crucial for fabricating long length MgB_2 tapes or wires with high critical current, which leaves Fe and Ni the ideal candidates [4][5]. Grasso et al. [6] first reported high critical current density of 10^5 A/cm^2 at 4.2 K on nickel-sheathed tapes without any heat treatment. High J_c of about 3 $\times 10^5$ A/cm^2 was acquired by extrapolating the $J_c - B$ curves to zero field. Wang et al. [9][10] investigated the Fe-clad MgB_2 tapes and wires in the temperature regime above 20 K and obtained J_c of 10^4 A/cm^2 at 30 K and 1 T. Komori et al. [11] reported J_c of the MgB_2 tape exceeding 10^5 A/cm^2 at 4.2 K and 10 T. However, all these results were obtained on short samples of several centimeters long. For large-scale applications, it is necessary to manufacture long length MgB_2 wires, tapes as well as coils with large critical currents.

Grain refinement can be achieved using finer starting powder, increased degree of powder pulverization during wire and tape fabrication, or by incorporating nano-scale, chemically inert particles which inhibit grain growth. In this paper, we first report on the improvement of the transport critical current density in iron-clad MgB_2 tapes by using ultra-fine starting precursor and the standard powder-in-tube process. We then scale-up this process to manufacture meter long MgB_2 wires. Transport critical current measurements on a solenoid wound from long wire are also presented.

EXPERIMENTAL PROCEDURES

The powder-in-tube method was used to fabricate the iron-clad MgB_2 wires and tapes. The low-cost 1020 iron-tube (plain carbon steel with 0.2% of carbon) had an outer diameter of 6 mm and a wall thickness of 1 mm. We crimp to seal one end of the iron tube, and then fill the tube with the desired precursor powder. The remaining end of the tube is crimped afterwards. Commercially available Mg powder (Alfa-Aesar, nominally 99.8% pure, -325 mesh) and B powder (Alfa-Aesar, nominally 99.99%, amorphous phase, -325 mesh) are stoichiometrically mixed. This powder mixture is then milled by Spex-8000 high-energy ball mill for 2 hours. Stainless steel balls and vial are used as milling medium and the mass ratio of ball to powder was 20:1.

The entire filling procedure was carried out in an argon atmosphere. The packing density was about 1.5 g/cm^3. We rolled the powder-filled tube using a groove rolling mill. A square wire with the dimension of 1 mm by 1 mm in cross-section area was obtained first and then flat-rolled to a tape with about 2.5 mm by 0.2 mm in cross-section area. The as-rolled tapes were very flexible and the surface was metal-shining as showed in Fig. 1. These tapes were wrapped by iron foil and annealed at 850 °C for 30 minutes. A high purity argon gas flow was maintained throughout the heat treatment process. The mass change of the tape after annealing was less than 0.2 %. The cross-section area of the MgB_2 core was

Fabrication of High Temperature Superconductc

1.07×10^{-3} cm^2, whereas the superconducting fill factor was about 20% of the whole conductor cross-section area (Fig. 2).

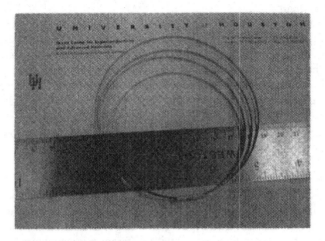

Figure 1. A meter long as-rolled flexible iron-clad MgB$_2$ tape.

Figure 2. Cross-section area of the annealed MgB$_2$ tape.

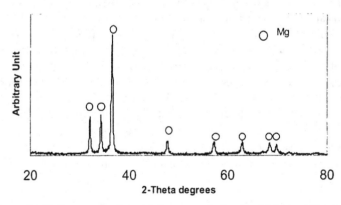

Figure 3. XRD pattern of the high-energy ball milled Mg and B mixture powder.

RESULTS AND DISCUSSIONS

Starting Precursor

Fig. 3 shows the x-ray diffraction pattern of the as-milled Mg and B mixture powder. All the XRD peaks are associated with Mg, which indicates that within 2 hours of milling, Mg and B do not react to form MgB_2 phase. It is not surprising that no B peaks are observed since B powder is in the amorphous phase. The broadening of the x-ray peaks reveals that 2-hour ball mill decreases the Mg particle size, which is confirmed by the SEM. The initial particle size of Mg and B were all about 40 μm, whereas after high-energy ball milling, the particle sizes were reduced to 200 to 300 nm. No iron contamination from milling medium could be detected.

MgB_2 Tapes

Fig. 4 shows the SEM photograph of the annealed MgB_2 tape along the longitude direction. A very uniform MgB_2 core can be seen in the middle without obvious pores or cavities. The MgB_2 grains align parallel to its axial direction. The iron sheath includes three layers with different microstructure features. The layer connected to the MgB_2 core is a loose-iron layer, which is due to the penetration of core materials. Outside this layer is a dark layer, which is composed of carbon and iron as indicated by EDS. Outside the carbon-rich iron layer is the dense-iron layer.

The critical temperature and critical current of the iron-clad MgB_2 tapes were measured by the standard four-probe method. A 20 mm long piece was cut from the tape and directly soldered to the current and voltage pads. A pulse current source with the maximum current of 225 A was used. A magnetic field was applied perpendicular to the tape length when the field dependence was measured. The voltage criterion was 1 μV/cm. Fig. 5 shows the resistance-temperature curve of the annealed iron-clad MgB_2 tape. The superconducting transition starts at 38.2 K and ends at 36.1 K, which gives a transition region of 2.1 K.

Fig. 6 shows the field dependence of critical current density on the annealed iron-clad MgB_2 tape. Solid lines are the measured data, whereas dotted lines are the extrapolated results. The critical current density is 2.0×10^5 A/cm^2 (critical current is 215 A) at 20 K under 0.6 T, and 1.1×10^5 A/cm^2 (critical current is 115 A) under 1.5 T magnetic field. Due to the limitation of our current source, we could not obtain the critical current at 20 K in self-field. However the extrapolation of $J_c - B$ curve gives J_c of 3.0×10^5 A/cm^2 at 20 K and zero-field. This tape also shows an impressive J_c of 1.6×10^5 A/cm^2 under 0.6 T and 5.6×10^4 A/cm^2 under 1.5 T at 25 K. The J_c value at 30 K drops to 6.5×10^3 A/cm^2 under 1.5 T.

Fabrication of High Temperature Superconducto

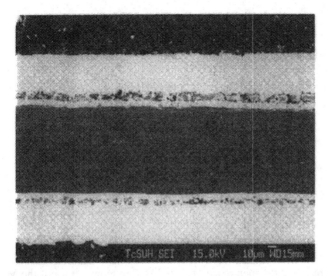

Figure 4. SEM photograph of the longitude section of the MgB$_2$ tape.

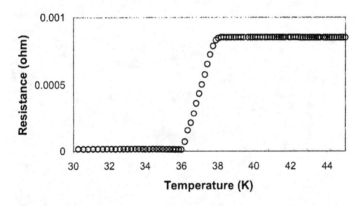

Figure 5. Temperature-resistance dependence of the MgB$_2$ tape.

Long MgB$_2$ Wires and Solenoid
  For large-scale applications, it is essential to fabricate long-length MgB$_2$ wires and tapes with the comparable transport critical current to short-length samples. In order to evaluate the scale-up capability of the wire/tape fabrication process, a 10-meter long MgB$_2$ square wire with the cross-section area of 1 mm by 1 mm was manufactured.

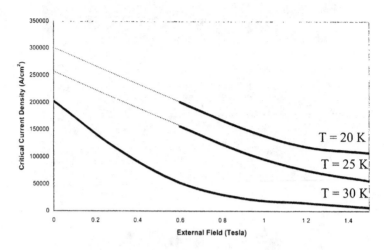

Figure 6. Field dependence of J_c at different temperature of the MgB$_2$ tape.

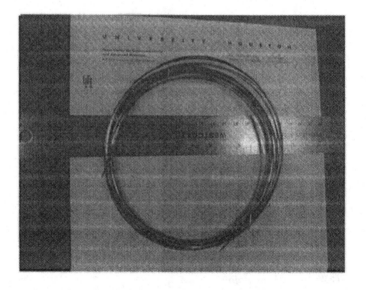

Figure 7. A 10-meter long as-rolled MgB$_2$ square wire.

Fabrication of High Temperature Superconducto

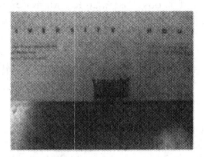

Figure 8. A solenoid of 12 mm diameter and 12 turns.

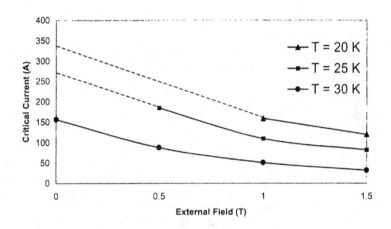

Figure 9. Field dependence of critical current at different temperature of the solenoid.

This long square wire was very uniform and flexible as shown in Fig.7. No open cracks and leakage of MgB$_2$ core material could be perceived. We further wound this green wire along a bolt of 12 mm diameter to form coils. A solenoid with 12 mm diameter and 12 turns prepared using this winding procedure is shown in Fig. 8. The solenoid's dimension was confined by the spatial limit of our testing facility. The solenoid was sintered at 850 °C for 30 minutes under a high purity argon gas flow. Current and voltage contacts were soldered to the solenoid directly. Two voltage contacts were 6 turns (23 cm) away from each other. Fig. 9 shows the field dependence of the critical current at different temperatures of the annealed solenoid. Solid and dotted lines represent measured and extrapolated results, respectively. Due to the limitation of our current source, no transition could be observed at 20 K until more than 0.5 Tesla external field was applied. At 1 Tesla, the critical current is 159 A at 20 K, 110 A at 25 K, and 51 A at 30 K. Even at 1.5 Tesla, this solenoid still possesses an impressive critical current value of 120 A at 20 K. The extrapolation to self-field of this critical current ~ external field curve gives a critical current of 340 A at 20 K and 270 A at 25 K.

CONCLUSIONS

We fabricated superconducting iron-clad MgB_2 wires and tapes via the standard powder-in-tube process using an ultra-fined Mg and B mixture precursor prepared by high-energy ball milling. The critical current density of the MgB_2 tape is 2.0×10^5 A/cm^2 at 20 K at 0.6 T, and 1.1×10^5 A/cm^2 at 1.5 T magnetic field. The critical current of a solenoid wound from MgB_2 wires is 159 A at 20 K and 1 Tesla, while the extrapolated critical current is 340 A at 20 K and self-field. These results give promise to applications of MgB_2 tapes in the temperature range of 20 ~ 30 K.

ACKNOWLEDGEMENT

This work was supported by the State of Texas through the Texas Center for Superconductivity and Advanced Materials (TcSAM), University of Houston.

REFERENCES

[1] J. Nagamatsu, N. Nakagawa, T. Muranaka, Y. Zenitani, and J. Akimitsu, Nature, **410**, 63 (2001).

[2] D. C. Larbalestier, L. D. Cooley, M. O. Riikel, A. A. Polyanskii, J. Jiang, S. Patnaik, X. Y. Cal, D. M. Feldmann, A. Gurevich, A. A. Squitieri, M. T. Naus, C. B. Eom, E. E. Hellstrom, R. J. Cava, K. A. Regan, N. Rogado, M. A. Hayward, T. He, J. S. Slusky, P. Khalifah, K. Inumaru, and M. Hass, Nature, **410**, 186 (2001).

[3] P. Grant, Mat. Res. Soc. Symp. Proc., **689**, 3 (2001).

[4] S. Jin, H. Mavoori, C. Bower, and R. B. van Dover, Nature, **411**, 563 (2001).

[5] S. Jin, H. Mavoori, C. Bower, and R. B. van Dover, Mat. Res. Soc. Symp. Proc., **689**, 11 (2001).

[6] G. Grasso, A.Malagoli, C. Ferdeghini, S. Roncallo, V. Braccini, and A. S. Siri, Applied physics letters, **79**, 230 (2001).

[7] H. Kumakura, A. Matsumoto, H. Fujii, and K. Togano, Applied Physics Letters, **79**, 2435 (2001).

[8] H. Suo, C. Beneduce, M. Dhalle, N. Musolino, J. -Y. Genoud, and R. Flukiger, Applied Physics Letters, **79**, 3116 (2001) .

[9] X. L. Wang, S Soltanian, J. Horvat, A. H. Liu, M. J. Qin, H. K. Liu, and S. X. Dou, Physica C, **361**, 149 (2001).

[10] J. Horvat, X. L. Wang, S. Soltanian, and S. X. Dou, Applied Physics Letters, **80**, 829 (2002).

[11] K. Komori, K. Kawagishi, Y. Takano, H. Fujii, S. Arisawa, H. Kumakura, M. Fukutomi, and K. Togano, Applied Physics Letters, **81**, 1047 (2002).

Fabrication of High Temperature Superconducto

FLUX LOSS MEASUREMENTS OF Ag-SHEATHED Bi-2223 TAPES

Mi-Hye Jang, W.Wong-Ng, R. Shull, L.P.Cook, DeaSik Suh,* and Taekuk Ko*
MSEL National Institute of Standards and Technology
100 Bureau Drive Stop 8522 Gaithersburg, MD 20899,USA
* Electrical Department, Yonsei University, Seoul, Korea

ABSTRACT

Alternating current (AC) losses of two Bi-2223 ([Bi, Pb] : Sr : Ca : Cu :O = 2:2:2:3) tapes [(Tape I, twist-pitch of 70 mm) and the other with a twist-pitch of 8mm (Tape II)] were measured and compared. These samples, produced by the powder-in-(Ag)tube (PIT) method, are multi-filamentary. Susceptibility measurements were conducted while cooling in a magnetic field. Flux loss measurements were conducted as a function of ramping rate, frequency and field direction. The AC flux loss increases as the twist-pitch of the tapes decreased, in agreement with the Norris Equation.

INTRODUCTION

Recent achievements in the fabrication of long-length multi-filament (Bi,Pb)-Sr-Ca-Cu-O (BSCCO) high-temperature superconductor (HTS,Type II) tapes with high critical current have generated considerable interest in applications such as cables, transformers, motors and generators, and energy storage systems. Since BSCCO tapes in most large scale systems are exposed to time-varying fields (or transport alternating currents), the tapes exhibit energy dissipation mainly due to AC losses. Thus, much research has been directed toward understanding of the nature and minimization of these AC losses [1-5]. In general, AC losses are dependent on the geometry of the filaments in the tapes, the magnetic and electrical properties of the superconductor, the type of matrix materials, and the amplitude and frequency of the transport current. The total losses, Q, are attributed to hysteretic loss within the filaments (Q_h), eddy current loss and coupling current loss within the matrix (Q_e), and coupling current loss across the matrix (Q_c). The Critical State model has been used extensively to describe the electrodynamics of Type II superconductors and to calculate the AC hysteresis loss [6-7]. For Bi-2223 ([Bi,Pb]:Sr:Ca:Cu:O=2:2:2:3)-based tapes consisting of multiple superconducting cores in a silver matrix, AC loss was found to be mostly frequency independent [6-8].

In this paper, the AC losses of two BSCCO tapes were measured by controlling the field magnitude and the incident angles between the c-axis of the tape and the time-varying magnetic field (60Hz). The influence of interior structure (i.e. different twist-pitch) on the transport current and AC loss was also determined.

EXPERIMENTAL

(1) Manufacturing of multi-filament silver-sheathed Bi-2223/Ag tapes

Multi-filament silver-sheathed $Bi_{1.8}Pb_{0.4}Sr_{2.0}Ca_{2.2}Cu_{3.0}O_{10+x}$ (Bi-2223/Ag) tapes were prepared by a powder–in-tube (PIT) technique where a high c-axis grain alignment is achieved by a combination of pressing, rolling and heating [9]. To prepare the Bi-2223 sample, appropriate amount of Bi_2O_3, $SrCO_3$, $CaCO_3$ and CuO were mixed and milled for 24h in methanol with ZrO_2 media. The milled slurry was dried and then calcined at 700 °C for 12h, 800°C for 8h, 835°C for 8h and 855°C for 8h. The calcinations and grinding procedures were repeated three times. To prepare the Ag -sheathed filaments, BSCCO powder was loaded into silver tubings followed by repeated rolling, drawing and heating. A total of 37 filaments were placed inside another Ag-sheathed tubing (6.35 mm outer diameter, 4.35mm inner diameter) and pressed. Heat treatment conditions are listed in Table 1. A final additional heat treatment in the absence of oxygen was also used to ensure complete sintering and reaction of the precursor powders. Microstructural observations were made to evaluate the uniformity of the superconductor filaments that were deformed during the twisting process.

Figure 1 shows a schematic diagram of the apparatus used for twisting the Bi-2223/Ag multi-filament tapes. The speed of drawing (RPM) was usually between 32~2000 RPM depending on the length of the twist-pitch. During twisting, the sample was further rolled and annealed twice at 850 °C for 150h. The final multifilament Bi-2223 tapes are both 37-core with Ag matrix. Tape I has a twist-pitch of 70mm and Tape II has a twist-pitch of 8 mm. The specifications of Tape I and Tape II are given in Table 2.

The microstructure of these tapes was evaluated by optical and scanning electron microscopy (SEM) on both the polished and fractured surfaces. The degree of texturing, the twist pitch, and the presence of cracks were determined after etching the Ag sheath with a mixture of $H_2O_2 : NH_3 = 1:1$.

Fabrication of High Temperature Superconductor

Table 1. Heat treatment conditions of the BSCCO powder

Parameters	Value
Heating rate	5 °C /min
Annealing time	850 °C/150h
Atmosphere	O_2 gas
Sintering rate	3.33 ° /min

Table 2. Characteristics of the two multi-filamentary Bi-2223/Ag tapes (37-core)

	Characteristics	
	Tape I	Tape II
Width	2.29 mm	2.29 mm
Thickness	0.15 mm	0.15 mm
Length	6.23 mm	6.23 mm
Weight	34.05 mg	34.05 mg
Twist pitch	70 mm	8 mm
Matrix	Ag	Ag
Density	15.9 mg/mm^3	15.9 mg/mm^3
Filling factor	2.2	2.2

(2) Magnetic Flux Loss Measurements

AC magnetic susceptibility was measured using a computer-controlled AC magnetometer (ACM) with AC voltage applied at a frequency of 125 Hz (harmonic 1), and external magnetic field of 10 Oe. The measurements were carried out as a function of temperature from 4.2 K to 130 K. AC losses were determined by using data measured in a superconducting quantum interference device (SQUID) magnetometer and in an AC/DC magnetometer (ACM/DCM).

Figure 2 shows the schematic drawing of the main part of the AC magnetometer. Magnetization within the sample can be determined by integrating the e.m.f. generated in the pickup coil. In these experiments, only output data from the current source of the pickup coil was used. The voltage generated by an external time-varying source was used to study the AC losses. All signals from the lock-in amplifier and data acquisition device were stored in the analog recorder.

The samples were first cooled to the lowest temperature in zero magnetic field. The measurements were then performed in a magnetic field during warming. A change of magnetization with time while the magnetic field was applied can be explained by thermo-activated movements of vortices through the maze of pinning centers.

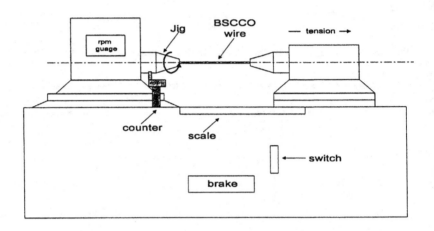

Fig. 1. Schematic drawing of the experimental apparatus used to prepared the PIT tapes

Two types of experiments were performed to measure the flux loss. In the first type, the critical current of each sample was determined at 4.2K, 60K and 77K from the peak magnetization field in the hysteresis loops. The magnitude of the external field and its incident angles to the c-axis of the sample were measurement variables. In the second type of experiment, AC loss measurements were carried out under constant transport current and time-varying external field of 200 Oe rms, 1000 Hz.

Fabrication of High Temperature Superconducto

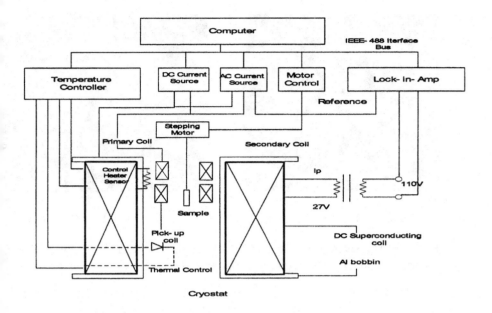

Fig. 2. Schematic drawing of the main part of the experimental apparatus used to measure AC loss

RESULTS AND DISCUSSION

Figure 3 shows the SEM micrographs of a typical untwisted tape (twist-pitch = ∞ mm) and a typical twisted tape, both after Ag-sheath etching (25 times). Exposed filaments of Tape I were uniformly deformed and their interfaces were also uniform and straight. Exposed filaments of the 8mm-twisted Tape II wire were rotated 55 turns, and also retained uniformity throughout the entire length.

Figure 4 shows hysteresis loops measured at 4.3K, 60K and 77K for Tapes I and II. For Tape I (large twist-pitch), an initial small field (-10 Oe ≤ H ≤ + 10 Oe) loop (not shown) indicated no magnetic flux penetration into the sample. An initial slope of -0.078 emu/cm^3 Oe is very close to the theoretical value of $-1/(4 \pi)$ for complete diamagnetism). Flux penetration (deviation from linearity of M vs. H) in Tape I at 4.3K begins to occur at a magnetic field near 30 Oe (= H_{c1}) (too small to show).

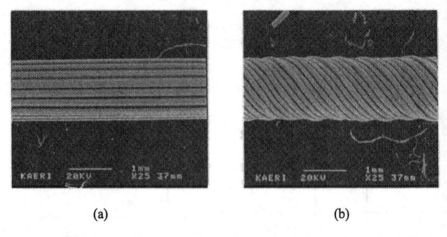

(a)   (b)

Fig. 3. SEM images of (a) A typical untwisted tape (twist-pitch = ∞ mm) (b) A typical twisted tape, both after Ag-sheath etching (25 times)

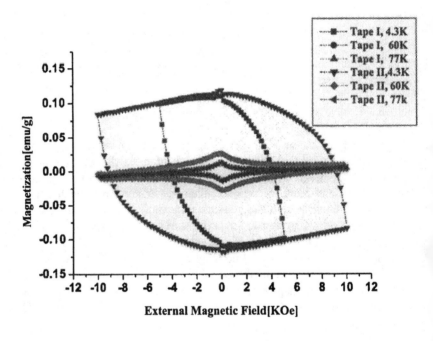

Fig. 4. Temperature dependence hysteresis loops of sheathed $Bi_{1.8}Pb_{0.4}Sr_{2.0}Ca$ $Cu_{3.0}O_{10+x}$ (Bi-2223/Ag) Tapes I and II, with virgin curve at 10 KOe

For Tape II, the initial small field (-20 Oe ≤ H ≤ +20 Oe) loop (not shown) also indicated no magnetic flux penetration. Its significantly smaller initial slope of 0.002 as compared to that of Tape I, was only 0.24 times the theoretical value of $-1/(4 \pi)$ for complete diamagnetism. Flux penetration at 4.3 K began to occur in this sample at a higher field of 68 Oe (=Hc_1).

From the large width of the loops for both tapes as shown in Fig. 4 (-5000 Oe ≤ H ≤ +5000 Oe for Tape I, and -10,000 Oe ≤ H ≤ +10,000 Oe for tape II), flux pinning is significant in these samples at 4.3 K. Magnetization of both Tapes I and II taken at 60K, are similar to each other, as are also the results at 77K. However much less flux pinning is observed at these elevated temperatures.

Based on equations (1) to (3) as shown below [10], AC loss is essentially due to magnetization variation with applied field at varying temperature, and varying applied field.

$$Q = \frac{1}{8} * (\frac{a}{b})^2 (\frac{Lp^2}{2*10^9}) Hm * \frac{dH}{dt} \tag{1}$$

$$Q = \frac{1}{8} * (\frac{a}{b})^2 (\frac{Lp^2}{2*10^9}) \frac{Hm}{\Delta \frac{\Delta Q}{dH}{dt}} \tag{2}$$

and

$$dH/dt = \frac{5*[M]*10^4}{time\ (min)*60}, \quad M = 0.1, 0.5, \text{ and } 1.7\ kOe \tag{3}$$

where, dH/dT is the time derivation of the the magnetic field, Q is the total loss, Lp is the twist pitch of core, V, a and b are the volume, width and length of the core, respectively, Hm is the maximum magnetic field.

The AC losses of Tape I and Tape II as a function of magnetic field direction (incident angle with respect to the c-axis of the tape) and ramping rate under an external magnetic field of 1 Tesla are illustrated in Fig. 5. As expected, the AC losses at an incident angle of 0 ° were lower than those at an angle of 90 °. The critical current of Tape II was slightly lower than that of Tape I, although the difference is too small to be observed on the same plot scale. The data in Fig. 5 also include the self-hysteresis loss due to the transport current. When the field is parallel to the c-axis of the tape, the measured loss was significantly smaller than that when the field is perpendicular to the field. The AC losses of Tape I (90 ° inclination) are about 64% of that of Tape II. This phenomenon is probably due

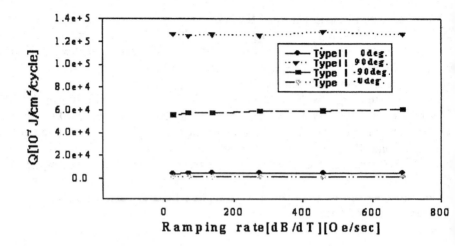

Fig. 5. AC Losses as a function of magnetic field direction and ramping rate for Tape I (70 mm twist-pitch) and Tape II (8mm twist-pitch) tape, both under an external magnetic Field of 1 T.

to the lower critical current of Tape II as compared to Tape I. For the C inclination experiments, both Tapes I and II have similar AC loss magnitud which may be due to similar flux areas in both tapes.

Figure 6 shows the AC loss of Tape I as a function of frequency and als with different incident angle to the c-axis of the tape. These data include the sel hysteresis loss due to the transport current. When the applied field was parallel the c-axis of the tape at 77K, the measured loss was significantly smaller tha that when the field was perpendicular to the c-axis. When the c-axis parall field was applied at 60K and 4.2K the measured loss was also smaller than with perpendicular field. It is not clear at this time why the samples experience th same loss when the field was applied at 60K and at 4.2K (under both parallel an perpendicular conditions).

It has been suggested that the AC losses of filament BSCCO tape, in genera will be effectively reduced when the filament size, electrical resistivity of the matrix, and the twist pitch of the filaments are properly modified [11]. The relation between AC losses and twist pitch can be expressed by the following equations [8]:

Fabrication of High Temperature Superconductc

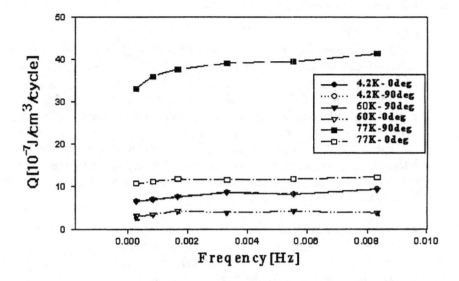

Figure 6. AC losses of Tape I as a function of frequency (according to temperature and incident angle with the c-axis). Data overlaps are shown for datasets collected at 4.2 K and 60K at both incident angles of 0°and 90°.

$$Qc = \frac{n\,\pi B a^2\,\omega\tau}{\mu_o(1+\omega^2\tau^2)} \qquad \text{(Norris Equation)} \qquad (4)$$

and

$$\tau = \mu_o\sigma_c Lp^2\,\frac{d^2{}_c}{16\omega^2{}_c} \qquad (5)$$

where Qc is the coupling current loss, n is the sharp factor, B_a is the external field amplitude, ω is the angular frequency of the field, τ is a time constant, μ is initial permeability, σ_c is the conductivity of the matrix, L_p is the twist pitch, and d_c and ω_c are the thickness and width of the type core, respectively. According to the Norris Equation (Eq. 4), it is expected that the AC losses will increase as the twist pitch decreases. Our present observations agree with the prediction of this equation. The decrease in twist pitch will also likely give rise to microstructural damage.

CONCLUSION

We have compared the loss factors of two BSCCO-tapes with two different twist pitches and two different incident angles of the magnetic field to the c-axis of the tape. AC losses for the textured Bi-2223 tapes are significantly higher when the magnetic field is applied perpendicularly to the c-axis than when it is applied in parallel direction. Also, the AC loss found in Tape I which has a large twist-pitch of 70 mm (essentially an untwisted tape) is approximately 64% of that of Tape II (twist-pitch of 8 mm). The larger loss in Tape II may be related to interface irregularity, smaller grain size, poor texture, and the presence of cracks caused by induced strain during the twisting process. We will continue to investigate the cause of these differences.

ACKOWLEDGEMENTS

This work was supported by the Post-doctoral Fellowship Program of the Korea Science & Engineering Foundation (KOSEF). Rosetta Drew of NIST is thanked for her assistance in the flux loss measurements.

REFERENCES
1. J. Yoo, J. Ko, H. Kim and H. Chung, IEEE, Trans. Appl. Supercond. 9 [2] 2163 2166 (1999).
2. Y. Yang, T.J. Hughes, C. Beduz, F. Durmann, Physica C 310 147-153 (1998)
3. F.Darmann, R.Zhao, G. Mccaughey, M.Apperley, and T.P.Beals, and C. Friend, IEEE, Trans. Appl. Supercond. 9 [2] 789-792 (1999).
4. W.Goldacker, H.Eckelmann, M.Quilitz, B. Ullmann, IEEE, Trans. Appl. Supercond. 7 [2] 1670-1673 (1997).
5. J. R. Cave, A Fevrier, T. Verhaege, A. Lacaze, Y. Laumond, "Reduction of AC losses in ultra-fine multi-filamentary Nb-Ti wires," IEEE Trans. Magnet. 25 1945-1948 (1989).
6. M. Polak, I. Hlasnik, S. Fukui, N. Ikeda, O. Tsukamoto "Self-field effect and current-voltage characteristics of AC superconductors," Cryogenics 34 315-324 (1994)
7. W. Swan, J. Math. Phys. 9 1308 (1968).
8. W. T. Norris,"Calculation of hysteresis losses in hard superconductors carrying ac: isolated conductors and edges of thin sheets," J. Phys. D 3 489 (1970).
9. E. H. Brandt, M. Indenbom. Phys. Rev. B48 12893-12906 (1993).
10. W. J. Carr, Jr. *AC loss and Macroscopic theory of superconductors*, pp. 86, published by Gordon and Breach, New York, 2001.

11. M. Jaime, M. N. Regueiro, M. A.A.Franco, C. Chaillout, J.J. Capponi, A. Salpice, J. L. Tholence, S. de Brion, P. Bordet, M. Marezio, J. Chenavas, B. Souletie, Solid State Comm. **97** 131(1997).

Fabrication of High Temperature Superconduct

PREPARATION OF SrZrO₃ THIN FILMS ON Bi(2223) TAPES FOR THE REDUCTION IN AC LOSSES

Se-Jong Lee
Kyungsung University
110-1, Daejeon-Dong
Busan 608-736, Korea

Deuk Yong Lee
Daelim College of Technology
526-7, Bisan-Dong
Anyang 431-715, Korea

Yo-Seung Song and Kyung-Hwan Ye
Hankuk Aviation University
200-1, Hwajon-dong
Koyang 412-791, Korea

ABSTRACT

SrZrO₃ resistive oxide barrier on Ag sheathed Bi(2223) tapes prepared by the sol-gel and dip coating method was evaluated with an aid of Taguchi method and $L_{18}(2^1 \times 3^7)$ orthogonal arrays to determine the optimal process combination of levels of factors that best satisfy the bigger is better quality characteristic. For analysis of results, statistical calculations such as average and analysis of variance were employed to analyze the results for improving the performance qualities of the dip coated SrZrO₃ films. Experimentally, the performance of the films was evaluated in terms of bond strength by varying Sr/Zr mol ratio (A), amount of organic vehicle additives (B), drying temperature (C) and time (D), heat treatment temperature (E) and time (F), respectively. The optimal combination of levels of factors was determined to be $A_3B_2C_3D_2E_1F_3$ having a 90% confidence level. Also, no chemical reaction between SrZrO₃ coating and Bi(2223) was observed during fabrication of 19 multifilamentary Bi(2223) tubes.

INTRODUCTION

Bi(2223) superconducting multifilamentary tapes suffer alternative current (ac) losses caused by the magnetic interaction between two neighboring parallel superconducting tapes carrying a transport current because they are exposed to alternative currents (ac) and fields.[1,2] The use of the Ag sheath material enhances the considerable coupling currents flowing in the sheath under ac magnetic fields.[2] One of the methods for reducing ac losses is the introduction of resistive oxide barriers between the filaments.[3] The barrier should be an insulator, stable under Bi(2223) annealing conditions and easy to deform. Witz et al.[2] reported that $CaZrO_3$ and $PbZrO_3$ reacted with Bi,Pb(2223) during heat treatment, resulting in an excess of Ca or Pb and a deficiency of Sr inside the filaments. Similar behavior was observed for $CaWO_4$, $PbWO_4$, $CaMoO_4$ and $PbMoO_4$. They also argued that $SrTiO_3$ and commercial $BaZrO_3$ reduced the Bi,Pb(2223) formation rate, suggesting that those oxides were not adequate for the barrier. On the other hand, $SrZrO_3$ was proposed as a prospective oxide barrier because it had no influence on the Bi,Pb(2223) formation rate and no chemical reaction with the filaments.[2] $SrZrO_3$, which is commercially available and cheap, has been known to have high mechanical and chemical stability.[4]

$SrZrO_3$ resistive oxide barriers on Ag sheathed Bi(2223) tapes were prepared by the sol-gel and dip coating method to enhance adherence (bond strength) prior to the ac loss measurements. The quality engineering method developed by Dr. Taguchi has been applied to determine the optimal combination of levels of factors that best satisfy the bigger is better quality characteristic (QC=B). To meet the purpose of determining the design solutions, the use of signal-to-noise ratio (S/N) for analysis of repeated results, which is the mathematical formula used to calculate the design robustness, may give a sense of how close the design is to the optimum performance of a process. Experiments with 6 factors at 3 levels were accomplished using the $L_{18}(2^1 \times 3^7)$ orthogonal arrays to lay out experiments of particular factor constituents. For analysis of results, statistical calculation such as average and analysis of variance (ANOVA) were employed. In the present study, 6 three-level factors were considered as follows: (A) Sr/Zr mol ratio; (B) amount of ethyl cellulose; (C) drying temperature and (D) time; (E) heat treatment temperature and (F) time, respectively.

EXPERIMENTAL PROCEDURE

Ag sheathed Bi(2223) tape having a dimension of 20㎜×3㎜ was polished using a SiC grit of 600 and then surface treated using ultrasonic cleaner using distilled water. The precursor solutions were prepared from commercial reagent grade strontium acetate-hemihydrate ($Sr(CH_3COO)_2 \cdot 1/2H_2O$, Junsei, Japan) in glacial acetic acid by stirring and heating to 80°C. Zirconium(IV) propoxide (~70% in propanol, Aldrichi, Japan) was mixed with glacial acetic acid and then acetylacetone (99%, Junsei, Japan) was added.[5] Both solutions were mixed and stirred at 80°C and then the resulting solution was diluted with distilled water to form 0.5 M of $SrZrO_3$ precursor solution. Organic vehicle (ethyl cellulose) dissolved in α-terpineol (>95%, Kan, Japan), which was prepared at 60°C, was added to the $SrZrO_3$ precursor solution by weight. The solutions were mixed and stirred for 1 h at room temperature. Thin films were obtained from the precursor solution by successive dip coating on Ag sheathed Bi(2223) tapes. The films were dried for 5~10 min at 100~130°C and then heat treated for 10~20 min at 500~700°C to remove organic residues such as alkoxy groups and to form chemical bonds between film and Ag sheathed tape substrate.[5] This process was repeated successively up to 4 times.

Microstructure and phase of the film were observed by scanning electron microscopy (SEM, S-2400, Hitachi, Japan) and XRD (3000PTS, Seifert, Germany), respectively. Bond strength of the films was evaluated by a tape test according to ASTM D3359-95a.[6]

RESULTS AND DISCUSSION

Experimental bond strength having the properties of QC=B was converted to S/N ratio according to the equation of $S/N = -10 \times \log[1/n\Sigma(1/j^2)]$, where n and j are the degree of freedom and the bond strength, respectively, as listed in Tables I and II. S/N values having the same factor and level were summed up to examine the ranks of variables and analyze the S/N ranks instead of the original values.[5,7,8] The optimal experimental condition can be achieved when the S/N ratio becomes the largest among the experiments investigated. The optimal combinations of levels of factors were $A_3B_2C_1D_3E_1F_2$ (No. 17) and $A_3B_2C_3D_2E_1F_3$ (No. 8). The average and the contribution rate of individual levels were calculated based on the S/N ratio. The variation of the average values of levels increased as the experimental factors were varied from A→E→D→C→B→F, indicating that the influence of experimental factors on the bond strength became pronounced in the order of A(22.86)<E(5.17)<C(4.50)<B(4.44)<F(1.00). It suggested that the subdivided levels were directly related to the bond strength of the $SrZrO_3$ film, therefore, small deviation of the contribution rate (A, Sr/Zr mol ratio) having a higher value was more susceptible to the large divergence of the bond strength.

Table I. Experimental descriptions using level notations for $L_{18}(2^1 \times 3^7)$ and their S/N values

No	e	A	B	C	D	E	F	e	S/N
1	-	1	1	1	1	1	1	-	14.72
2	-	1	2	2	2	2	2	-	14.78
3	-	1	3	3	3	3	3	-	14.32
4	-	2	1	1	2	2	3	-	16.39
5	-	2	2	2	3	3	1	-	15.31
6	-	2	3	3	1	1	2	-	14.37
7	-	3	1	2	1	3	2	-	20.92
8	-	3	2	3	2	1	3	-	28.97
9	-	3	3	1	3	2	1	-	26.58
10	-	1	1	3	3	2	2	-	14.11
11	-	1	2	1	1	3	3	-	12.65
12	-	1	3	2	2	1	1	-	15.39
13	-	2	1	2	3	1	3	-	14.11
14	-	2	2	3	1	2	1	-	15.46
15	-	2	3	1	2	3	2	-	15.26
16	-	3	1	3	2	3	1	-	24.56
17	-	3	2	1	3	1	2	-	30.97
18	-	3	3	2	1	2	3	-	22.57
					Total sums				331.44

Square sums of S/N values divided by the number of experiments having the same factor and level (S), degree of freedom (φ) and mean squares, S/φ(=V), were determined. V value was incorporated into error term (pooling) because V value of factor F was the lowest among the experiments investigated, indicating that the contribution of F to the bond strength of the $SrZrO_3$ film may be insignificant. ANOVA analysis after pooling in listed in Table III. Ratio of variance (or F-ratio), $(V_C/V_e=F_o)$, was calculated and then compared with the theoretical $F(\varphi_e, \varphi_E; \alpha=0.1)$ value determined from the standard table in the ref. 9. The lowest F_o value of 9.59 (factor C) in the present study was examined in order to evaluate the difference with the values of $F(2,5; \alpha=0.1)$. F_o value was greater than that of $F(\varphi_e, \varphi_E; \alpha=0.1)$, implying that the drying temperature was a significant factor within $\alpha=10\%$. It is conceivable that the influence for the factors shown in Table III was significant with a 90% confidence level.

Table II. Factor and level descriptions

Level	e	A(mol%)	B(wt%)	C(°C)	D(min)	E(°C)	F(min)	e
1	-	0.7/0.3	3	100	5	500	10	-
2	-	0.5/0.5	5	130	10	600	15	-
3	-	0.3/0.7	7	160	15	700	20	-

Table III. ANOVA analysis after pooling

Factor	S	φ	V	Fo	F(0.1)
A	488	2	244	283.72	9.29
B	16.87	2	8.44	9.81	9.29
C	16.50	2	8.25	9.59	9.29
D	23.96	2	11.98	13.9	9.29
E	20.13	2	10.07	11.7	9.29
Error	4.31	5	0.86	-	
Total	569.77	15	284.94		

The top-coat film was analyzed using XRD to identify $SrZrO_3$ phase as shown n Fig. 1. Although the S/N ratio of No. 17 in Table I was the highest, no $SrZrO_3$ phase in Fig. 1 was observed, indicating that No. 17 specimen was not adequate or the formation of $SrZrO_3$ coating. Therefore, No. 8 specimen having the second ighest S/N ratio was determined to be the optimal combination of levels of factor $A_3B_2C_3D_2E_1F_3$) for the coatings.

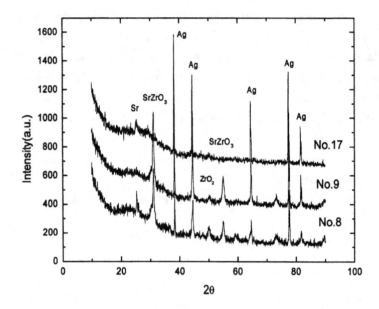

Fig. 1. XRD patterns of specimens.

The Ag sheathed Bi(2223) hexagonal tubes (14×12×300 mm^3) were dip coated for 4 times by using the optimized sol-gel precursor (No.8). Although the coatings were not thick enough to fabricate the multifilament, 19 coated Bi(2223) tubes were swaged for 5 times along with annealing for 1 h at 250°C before swaging as shown in Fig. 2, resulting in reduction in cross section area by 75% and a diameter of 18 mm. Thin dark SrZrO$_3$ interface between Ag sheaths was visible and no chemical reaction between SrZrO$_3$ and Bi(2223) was detected.

CONCLUSIONS

The optimum process condition of the SrZrO$_3$ film on Bi(2223) films for the bond strength was evaluated by Taguchi method and orthogonal arrays and determined by to be A$_3$B$_2$C$_3$D$_2$E$_1$F. Experimentally, the dip-coated SrZrO$_3$ film was composed of SrZrO$_3$ film having 0.3/0.7 Sr/Zr mol ratio, 5 wt% of ethyl cellulose, drying temperature and time of 160°C and 10 min, heat treatment

Fabrication of High Temperature Superconducto

temperature and time of 500°C and 20 min, respectively. The variation of the average values of levels increased as the experimental factors were varied from A→E→D→C→B→F, indicating that the influence of Sr/Zr mol ratio on the bond strength became the most pronounced. In conclusion, the influence for the factors was significant with a 90% confidence level.

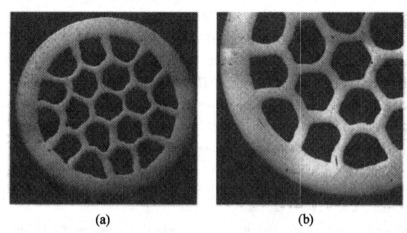

(a)   (b)

Fig. 2. Cross sectional view ((a) front and (b) back) of the 19 filament hexagonal tube with SrZrO$_3$ oxide barrier after 5 swaging. The final diameter of the multifilament tube was 18 mm.

ACKNOWLEDGEMENTS

This research was supported by a grant from Center for Applied Superconductivity Technology of the 21st Century Frontier R&D Program funded by the Ministry of Science and Technology, Republic of Korea.

REFERENCES

[1]M. Majoror, B.A. Glowacki and A.M. Campbell, "Transport ac Losses and Screening Properties of Bi-2223 Multifilamentary Tapes Covered with Magnetic Materials," *Physica C*, **338** 251-262 (2000).

[2]G. Witz, M Dhalle, R. Passerini, X-D. Su, Y.B. Huang, A. Erb and R. Flukiger, "AC Losses in Bi,Pb(2223) Barrier Tapes," *Cryogenics*, **41** 97-101 (2001).

[3]Y.B. Huang and R. Fulkiger, "Reducing ac Losses of Bi(2223) Multifilamentary Tapes by Oxide Barriers," *Physica C*, **294** 71-76 (1998).

[4]T. J. Eschenbaum, J. Rosenberger, R. Hempelmann, D. Nagengast and A. Weidinger, "Thin Films of Proton Conducting $SrZrO_3$ Ceramics Prepared by the Sol-gel Method," *Solid State Ionics*, **77** 222-225 (1995).

[5]S-J. Lee, D.Y. Lee, K-H. Ye and Y-S. Song, "Effect of Organic Vehicle Addition on Bond Strength of $SrZrO_3$ Thin Films on Bi(2223) Tapes," *Journal of Materials Science Letters*, **22** 315-318 (2003).

[6]Standard Test Methods for Measuring Adhesion by Tape Test. ASTM Designation D3359-95a, Annual Book of ASTM Standards, Vol. 06.01, 1996.2.

[7]K-S. Chae, H-K. Choi, J-H. Ahn, Y-S. Song and D.Y. Lee, "Application of Taguchi Method and Orthogonal Arrays for Characterization of Corrosion Rate of IrO_2-RuO_2 Film," *Journal of Materials Science*, **37** 3515-3520 (2002).

[8]K-S. Chae, H-K. Choi, J-H. Ahn, Y-S. Song and D.Y. Lee, "Effect of Organic Vehicle Addition on Service Lifetime of Ti/IrO_2-RuO_2 Electrodes," *Materials Letters*, **55** 211-216 (2002).

[9]R.K. Roy, "Experimental Design using Orthogonal Arrays:; pp. 95-135 in *Design of Experiments Using the Taguchi Approach*, John Wiley & Sons, New York, 2001.

NEW SEEDING METHOD FOR TEXTURING Y-BA-CU-O BULK SUPERCONDUCTOR: MULTIPLE SEEDED MELT GROWTH

Y.X.Zhou, H. Fang and K.Salama
Texas Center for Superconductivity
and Advanced Materials
University of Houston, TX 77204,
USA

U. Balachandran
Argonne National Laboratory
Argonne, IL 60439, USA

ABSTRACT

Top seeded melt process is a well known technique for manufacturing bulk superconducting discs with high trapped magnetic fields and levitation forces. This method, however requires a long time for sample growth and is limited in the length of samples grown (~10-15 mm). In this work, we have developed a seeding method, "Multiple Seeded Melt Growth (MSMG)", which reduces the manufacturing time and increases the length of samples. The basic idea is to press SmBCO seeds in the body of the green disk ($70\%YBa_2Cu_3O_7 + 30\%Y_2BaCuO_5$) before compaction. A modified heat treatment profile is then used to grow the bulk superconductor. The microstructure and crystal orientation were studied and found to be comparable to those of samples grown by top seeded melt growth method. We also studied the distribution of trapped magnetic fields in samples grown by this method. The results indicate that using this method, the sample is well oriented throughout the whole length in spite of decreasing the manufacturing time and increasing the sample length. These results give promise to the use of this technique in enhancing the superconducting properties of bulk textured YBCO.

INTRODUCTION

Tremendous efforts have been made in fabricating HTS superconductors in large grain form by a variety of melt processes. Melt processing has been successful in producing high Jc REBCO samples. Because these materials can sustain critical current densities (J_c) in the range of 10^4-10^5 A/cm^2 at 77K under external magnetic fields up to 3 T[1,2], and they are easy to reproduce in large batches[3], melt textured REBCO large grains have significant potential for a variety of high field permanent magnet applications such as generators, fly-wheel energy storage systems, and motors[4,5,6,7,8,9]. This, however, does not imply that the development of large grain REBCO has come to its conclusion. It is well known that top seeded melt process is the main technique for manufacturing bulk superconducting discs with high trapped magnetic fields, levitation forces and critical current densities. This method, however, requires a long time for crystal growth and is limited in the length of samples grown (~10-15 mm)[10,11,12]. When evaluating samples for practical applications, these parameters are very important and must be considered.

In order to reduce the manufacturing time and increase the length of samples, we have developed a seeding method, "Multiple Seeded Melt Growth (MSMG)". The microstructure and grain orientation were examined and the distribution of trapped magnetic fields and levitation forces throughout the sample were discussed.

EXPERIMENTAL

Various techniques for growing large grains have been developed based on the peritectic solidification process and a different philosophy of improving the performance of the product. In this experiment, the precursor powders were prepared by the solid-state route. Y-123, Y-211, and Pt powders were mixed according to the weight ratio 70:30:0.5 using dry milling. The mixed powders were then pressed into disks of 32 mm in diameter and 20 mm or 30 mm in thickness. The 211 disks were pressed as the support of 123. In order to study the Multiple Seeded Melt Growth effects, the SmBCO seeds were pressed into the raw disks and/or put on the top surface as shown in Fig. 1. We call them case-1, case-2 and case-3 respectively.

The prepared disks were then heated gradually in air in a box furnace to 1050°C and cooled at 0.5°C/hour to 990°C followed by furnace cooling to room temperature as shown in Fig. 2. During this process, the precursor body is melted at a temperature lower than the melting point of the seed followed by solidification of the precursor body so as to grow large REBCO grains. These large grains were further annealed in oxygen at temperatures between 350 - 450°C for 100 hours in order to develop the superconducting properties. A typical heating process is shown in Fig.2. In the traditional top seeded melt growth method, researchers usually set the end cooling temperature at 950 °C and in some cases even lower. Therefore, the manufacturing time is reduced significantly.

In order to study the orientation of sample along the axial direction, the disks are sliced parallel to the surface and characterized by X-ray diffraction.

The distribution of the normal component of the trapped magnetic field at 77 K on the surface of each disk was measured by scanning a Hall probe at a fixed distance (about 0.5 mm) from the top surface of the specimens. The magnetic fields were trapped by cooling the specimens to liquid nitrogen in an external field of 1.5 T.

Fabrication of High Temperature Superconducto

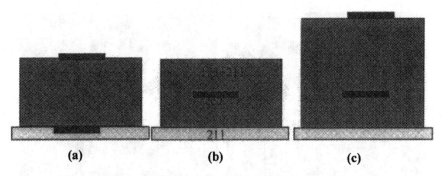

(a)	(b)	(c)

Fig. 1. Schematic of three types of seeding techniques applied:
(a) Double Seeds Case-1. (b) Single Middle Seed Case-2. (c) Top-Middle seeds Case-3.

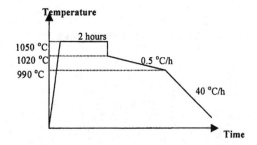

Fig.2. Temperature program Chart of MSMG of YBCO Large Grain.

RESULTS AND DISCUSSIONS

For a sample seeded as Case-1 of Fig. 1, it was found that both top and bottom parts of the sample are well oriented and the superconducting properties in both halves are almost the same. In other words, the superconducting properties of the sample in bottom part are improved significantly. We already published these results[12], and therefore, this case will not be discussed here.

Fig. 3 shows photographs for the top view and side view of a bulk sample of YBCO large grain fabricated as Case-2 of Fig. 1. It can be seen that the growth morphology shown in Fig. 3. (a) exhibits a striking feature: a rectangular facet plane forms on the top surface due to the accelerated growth rate of the grain along the a and b axes of the lattice. The resulting facets, therefore, determine the a and b axes of the fully grown YBCO grain. A well defined line from top to bottom in the side of the sample is shown in Fig. 3 (b), which is formed also due to the accelerated growth rate of the grain along the a and b axes of the lattice. Because the top and bottom part form one straight line and normal to the surface, it indicates the whole sample along the length direction is well oriented.

(a)   (b)

Fig. 3. Optical micrographs of the bulk sample seeded in Case-2 of Fig. 1:
(a) top view and (b) side view.

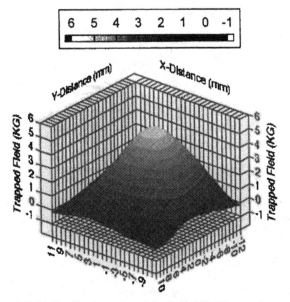

Fig. 4. Trapped field distribution map of the bulk YBCO sample seeded as Case
of Fig. 1.

The distribution of the normal component of the trapped magnetic field on the surface of YBCO grain fabricated in Case-2 of Fig. 1 is shown in Fig. 4. A maximum trapped field ($5.6KG$) is present in the center, as can be derived from the critical state of the magnetic flux lines in the grain. Due to the symmetry of the growth morphology about the center, a four-fold symmetry in magnetic field distribution is shown. Moreover, only a single peak is observed, indicating that weak links are absent in the sample and it is a single grain.

                Fabrication of High Temperature Superconduct

Fig. 5. Optical microstructure of the bulk YBCO sample seeded as Case-2 with c-axis orientation perpendicular to the top surface.

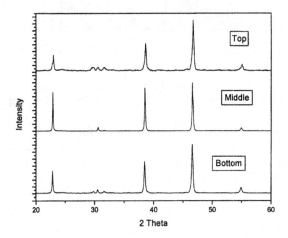

Fig. 6. The 2-Theta X-ray diffraction patterns of the bulk sample seeded in Case-2 of Fig. 1.

Melt processed samples characteristically contain platelet (well orientated (100), (010) and (001) thin planes) boundaries in planes perpendicular to the c-axis of the lattice that form during the peritectic growth process, as shown in Fig. 5. This figure shows the cross section of YBCO large grain fabricated in Case-2. In the center, there is a seed. It clearly shows that both the top and bottom parts are well textured and the ab planes are all parallel to the sample surface. Moreover, there is no large boundary close to the seed, which indicates both parts growing from the seed at the same time and forming a single crystal.

In order to study the orientation of the sample along the axial direction, the disks are sliced parallel to the surface and characterized by X-ray diffraction. Fig.

6 shows XRD patterns of the surface of the textured sample fabricated in Case-2. The strong (00*l*) peaks are observed indicating that the c-axis oriented 123

Fig. 7. Optical micrographs of the cross section for bulk sample Seeded in Case-3 of Fig. 1.

phase is grown along the grain axis of the Sm-123 seed crystal, and, all the disks consist of only intense (00*l*) peaks, which indicate the whole sample has c-axis orientation.

Based on the case-1 and Case-2, we successfully fabricated 25 mm long YBCO large grains using the Case-3 technique. Fig.7. shows the optical micrographs of the cross section for the sample. It is clearly shown that the platelet boundaries formed during the peritectic growth process along the whole sample are perpendicular to the c-axis. In other words, the whole sample is well oriented.

CONCLUSIONS

In this work, we have developed a seeding method, "Multiple Seeded Melt Growth (MSMG)", by which large domain YBCO levitators were successfully fabricated. With this approach, the manufacturing time is reduced and the length of the sample can be increased significantly. The microstructure, crystal orientation, and superconducting properties were found to be comparable to those of samples grown by top seeded melt growth method. These results give promise to the use of this technique in enhancing the superconducting properties of bulk textured YBCO.

ACKNOWLEDGEMENTS

This work was supported by the Texas Center for Superconductivity and Advanced Materials and by the U.S. Department of Energy (DOE), Energy Efficiency and Renewable Energy, as part of a DOE program to develop electric power technology, under Contract W-31-109-Eng-38.

Fabrication of High Temperature Superconducto

REFERENCES

[1] W. Lo, D. A. Cardwell, C. D. Dewhurst and S. L. Dung, J. Mater. Res. 11 (1996) 786.

[2] M. Muralidhar and M. Murakami, Physica C 309 (1998) 43.

[3] D. Litzkendorf, T. Habisreuther, M. Wu, T. Strasser, M. Zeisberger, W. Gawalek, M. Helbig and P. Gornert, Mater. Sci. Eng. B53 (1998) 75.

[4] R. Decher, P.N. Peters, R.C. Sisk, E.W. Urban, M. Vlasse and D.K. Rao, Appl. Supercond. 1 (1993) 1265.

[5] W.K. Chu, K.B. Ma, C.K. McMichael and M.A. Lamb, Appl. Supercond. 1 (1993) 1259.

[6] T.A. Coombs, A.M. Campbell, I. Ganney, W. Lo, T. Twardowski and B. Dawson, Mater. Sci. Eng. B53 (1998) 225.

[7] L.K. Kovalev, K. V. Ilyushin, V. T. Perkin, K.L. Kovalev, Elect. Technol. 2 (1994) 145.

[8] I.G. Chen, J.X. Liu, Y. R. Ren, R. Weinstein, G. Kozlowski, O.K. Oberly, Appl. Phys. Lett. 62 (1993) 3366.

[9] G. Fuchs, G. Krabbes, P. Schatzle, P. Stoye, T. Staiger, K. H. Muller, Physica C 268 (1996) 115.

[10] K. Salama, A.S. Parih, L. Woolf, Appl. Phys. Lett. 68 (1993) 1996.

[11] Y.H. Zhang, A. Parikh and K. Salama, IEEE Applied Superconductivity 7(1997)1787.

[12] Yu X. Zhou, H. Fang and Kamel Salama *IEEE Trans. Appl. Supercond.* (in press).

Fabrication of High Temperature Superconduct

Control of Microstructure

ANALYTICAL TRANSMISSION ELECTRON MICROSCOPY OF THICK YBa$_2$Cu$_3$O$_{7-\delta}$ FILMS ON RABITS

Keith J. Leonard, Sukill Kang, Byeongwon Kang*, Amit Goyal and Donald M. Kroeger
Oak Ridge National Laboratory, Metals and Ceramics Division,
Superconducting Materials Group, One Bethel Valley Road
Oak Ridge, TN 37831-6116.

ABSTRACT
Microstructural changes related to the critical current density dependence on film thickness in superconducting YBa$_2$Cu$_3$O$_{7-\delta}$ (YBCO) films have been investigated through electron microscopy. Pulsed laser deposited YBCO films ranging in thickness from 0.19 to 6.4 μm were grown on rolling assisted biaxially textured substrates (RABiTS) having two different buffer layer architectures. Remarkable improvements in the YBCO microstructure were observed in samples deposited on Ni-3%W, with a thick Y$_2$O$_3$ seed layer. Films grown on this structure showed little random oriented grain formations along with minimal interfacial reactions. The causes of reduced performance with increased film thickness will be addressed for both RABiTS architectures.

INTRODUCTION
Difficulties in depositing or growing thick, high quality, c-axis oriented YBa$_2$Cu$_3$O$_{7-\delta}$ (YBCO) films whose properties do not become degraded with increasing film thickness, has been an elusive goal in producing commercially viable superconducting films. Typical problems in producing thick YBCO films can include: the development of misoriented YBCO grains or porosity,[1-5] formation of non-superconducting phases,[6-9] oxygen deficiencies[10] and changes in stoichiometry.[11] In addition, the choice of substrate can have a significant effect on the defects and microstructures generated within the YBCO film.[12] In this investigation, transmission electron microscopy was used to examine the causes for the reduction in J$_c$ with film thickness for pulsed laser deposited (PLD) YBCO

* Byeongwon Kang, is now with the National Creative Research Initiative Center for Superconductivity, Department of Physics, Pohang University of Science and Technology, Pohang, South Korea.

films on two different architectures of rolling assisted biaxially textured substrates (RABiTS). The architecture of the two substrates investigated were as follows:

RABiTS1:   Ni / CeO_2 / yttrium-stabilized zirconia (YSZ) / CeO_2 / YBCO
RABiTS2:   Ni-3at.%W / Y_2O_3 / YSZ / CeO_2 / YBCO

The use of Ni-3at.% W, hereby written as Ni-W, posses a sharper texture over earlier pure Ni tapes and is a recent improvement in the development of RABiTS materials.[13] The use of a Y_2O_3 seed layer instead of CeO_2 is the result of another upgrade in the RABiTS materials, due to concerns over crack formation in the CeO_2 seed layer and subsequent NiO growth into the upper layers of the coated conductor stack during processing.

EXPERIMENTAL

The 12 nm thick CeO_2 and 150 nm thick Y_2O_3 seed layers of the two substrates were deposited through electron beam evaporation of pure metals under a partial pressure of water onto the {100} <100> biaxially textured Ni and Ni-W substrates, respectively. Additional 150 to 200 nm thick YSZ and 20 nm thick CeO_2 layers were deposited by RF magnetron sputtering. Details in the processing conditions and substrate properties are covered elsewhere.[14-17]

YBCO films ranging in thickness from 0.19 to 6.4 μm were deposited by pulsed laser deposition (PLD) using a XeCl excimer laser (λ=308nm) on 2.5 cm x 0.5 cm samples. Deposition was done at 790°C under an oxygen partial pressure of 120 mTorr. Laser energy density was 4 J/cm^2 with a film growth rate between 5 to 13 Å/s. Following deposition, the films were cooled to room temperature at a rate of 5°C/min, under an oxygen partial pressure of 550 Torr.

Cross-section samples were prepared for transmission electron microscopy (TEM), either by tri-pod polishing or conventional ion-milling methods, details of which are described elsewhere.[9,18] The samples were investigated using a Philips Technai 20 (LaB_6, 200kV) and CM200 (FEG, 200kV) microscopes, equipped with energy dispersive spectrometry (EDS) units. The CM200 TEM was used in both conventional and high-resolution scanning modes.

RESULTS AND DISCUSSION

Properties of the Thick YBCO Films

From previous work,[16-17] the critical current densities (J_c) of the thick YBCO films along with the film textures determined through x-ray diffraction (XRD) are listed in Tables I and II for the two substrate types. The YBCO films characterized on RABiTS1 by TEM were 0.19, 0.5, 1.7 and 3.0 μm thick, and were similar in electrical and texture properties to the samples shown in Table I. The RABiTS2 samples characterized by TEM are those listed in Table II. The values of J_c were calculated from the four-point probe measurements of critical current (I_c) conducted at 77K in self-field without micro-bridge patterning. Limitations in characterizing the film properties to 120 amps prevented the full determination of I_c in films thicker than 1.0 μm of the RABiTS2 materials under self-field. For these films a zero-field J_c was calculated from in-field

Fabrication of High Temperature Superconducto

measurements at 0.5 Tesla, assuming that a factor of 4 to 5 drop in J_c occurs from self-field to that of 0.5 Tesla.[16] Therefore, a range of calculated values are presented for the 2.9 and 4.3 μm thick films in Table II. During measurement of the 6.4 μm thick film on RABiTS2, a crack was generated across the sample upon reaching 60 amps, believed to have been the result of sample heating.

Table I. Measured electrical and texture properties of thick YBCO films on the RABiTS1 material, from ref. [17].

YBCO thickness (μm)	J_c (MA/cm^2)	$\Delta\phi^*$	$\Delta\omega^{**}$	a-axis fraction $Ia_\perp(200) / Ic_\perp(002)$	% Cube
0.19	2.6	1.07	0.73	0.0	98.8
0.43	1.4	1.04	0.70	0.0	90.4
1.6	0.59	1.03	0.76	0.03	89.7
3.0	0.45	1.08	0.73	0.05	79.9

* In-plane texture: ratio of the FWHM of x-ray intensities of (113)YBCO to (111)Ni/Ni-W.
** Out-of-plane texture: ratio of (005)YBCO to (200)Ni/Ni-W.

Table II. Measured electrical and texture properties of thick YBCO films on the RABiTS2 material, from ref. [16].

YBCO thickness (μm)	J_c (MA/cm^2)	$\Delta\phi$	$\Delta\omega$	a-axis fraction $Ia_\perp(200) / Ic_\perp(002)$	% Cube
1.0	1.18	0.90	0.79	0.0	92.5
2.9	0.9 - 1.1	1.03	0.78	0.1	93.5
4.3	0.65 - 0.81	1.09	0.89	0.2	87.9
6.4		0.96	0.91	0.1	95.8

The fraction of a-axis ($a_\perp$) to c-axis ($c_\perp$) oriented grains within the films showed an increase with film thickness for both substrates, with that of the RABiTS2 materials being significantly higher. Although the larger $a_\perp$ concentration observed was substantially less than that appearing in YBCO films deposited on single crystal substrates under the same conditions.[17]

While the a-axis fraction of the Ni-W RABiTS2 showed a higher value than that of the Ni RABiTS1 substrates, the percent cube texture was significantly better despite the random variation in results measured. A slight increase in $\Delta\omega$ with thickness was also observed for the RABiTS2 materials.

Microstructural Changes with Film Thickness for RABiTS1.

The 0.19 μm YBCO film examined in cross-section (Fig. 1(a)) showed complete epitaxial c-axis orientation, with no secondary phase formation or porosity. The YBCO film consisted of c-axis oriented grain columns, separated by anti-phase boundaries whose contrast is visible due to the displacement of the (001) planes by a fraction of the unit cell dimension. Created either by ledge steps at the CeO_2 interface,[19] or by a variation in the initiating atomic stacking sequence of YBCO.[20] Deviations in $c_\perp$ columns across the anti-phase boundaries, measured as splitting of YBCO electron diffraction intensities, was negligible in the thinner films, but was found to increase with either film thickness or aging time.[9] Investigation of the YBCO/CeO_2 interface showed no evidence of reaction between the layers in the 0.19 μm film, nor the presence of any disordered perovskite layer within the YBCO as reported to occur in PLD and ex-situ grown films on single crystal and metal substrates.[20-21]

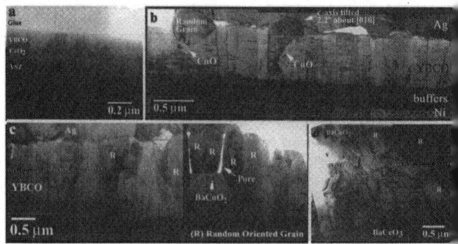

Fig. 1. Cross-section images of: (a) 0.19 μm, (b) 0.5 μm, (c) 1.7 μm, and (d) 3.0 μm thick YBCO films on RABiTS1.

Examination of the 0.5 μm thick YBCO sample on the RABiTS1 substrate (Fig. 1(b)) also showed $c_\perp$ columns extending the entire film thickness and no porosity. However, the film did show the formation of random oriented grains near the top surface beginning at a thickness of 0.45 μm, contributing to the drop in cube texture from 98.8% in the 0.19 μm film to 90.4%.

Observed within the 0.5 μm film was an outgrowth grain of YBCO, characterized as having an ~2.2° tilt in its c-axis, rotated about the [010] direction, relative to the surrounding film. The outgrowth originated 150 nm above the CeO_2 cap layer, and was not found to be associated with any defects within the buffer layer or the formation of secondary phases. A detailed explanation of the types of surface outgrowths observed in PLD films is provided elsewhere,[9] but it is believed to have occurred in this film as a result of a strain release associated the lattice mismatch between the growing YBCO film and CeO_2 layer.

As a result of reaction between YBCO and CeO_2 cap layer, $BaCeO_3$ particles of 10 to 20 nm in size were randomly observed along the interface (Fig. 2(a)).

Their formation occurred following the initial nucleation and growth of the YBCO, as no $BaCeO_3$ was observed in the 0.19 μm thick YBCO film. Due to the excess Y and Cu released from the formation of $BaCeO_3$, occasional particles of Y_2O_3 were observed within the film in addition to CuO formation along the interfaces between the misoriented and $c_\perp$ grains. Investigation of possible orientation relationships between CuO and YBCO revealed multiple growth directions possible for CuO.[9] It was found that while some CuO particles held a common relationship to the $c_\perp$ grains, others shared one with the random oriented grains. Suggesting that the formation of CuO followed random grain growth, and was not responsible for the misoriented YBCO grains.

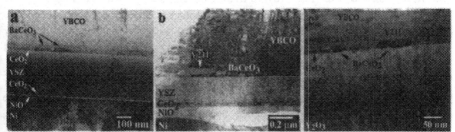

Fig. 2. Interfacial reactions between YBCO and CeO_2 in (a) 0.5 μm and (b) 3.0 μm thick YBCO films on RABiTS1, and in the (c) 6.4 μm thick RABiTS2 film.

The microstructure of the 1.7 μm film showed relatively undisturbed $c_\perp$ grains in the lower portion of the YBCO film (Figure 1(c)), however, the misoriented grains were more numerous near the top of the film surface. Unlike YBCO films grown by PLD on IBAD YSZ buffered substrates,[10] the misoriented grains do not completely cover the top of the film. The majority of the misoriented grains were formed in groups that typically shared a common twin boundary. The presence of porosity within the microstructures was also beginning to be seen, but was usually associated with the large mismatches between the $c_\perp$ and random oriented grains. These regions, however, may have contained secondary phases that were removed during sample preparation. CuO particles at the interfaces between the two types of grain orientations were also present, in addition to the formation of $BaCuO_2$.

The thickest YBCO film on RABiTS1, which suffered from the largest degradation in cube texture, showed a very rough, highly misaligned upper portion in its cross-section (Fig. 1(d)). Pores were found more frequently within the YBCO, which are believed to be associated with either the formation of secondary phases or the removal of quenched in vacancies.

Formation of CuO along the interfaces of the YBCO grains were again present within the 3.0 μm film, but a dramatic increase in the amount of $BaCuO_2$ was observed at locations both near the top surface of the film as well as at intermediate depths.

The extent of the reaction between the YBCO and CeO_2 cap layer appeared to be complete within the 3.0 μm sample (Fig. 2(b)). Investigation of the interface region through both diffraction and composition analysis could not identify any CeO_2 layer remaining. Y-enrichment resulting from the formation of $BaCeO_3$ did produce Y_2BaCuO_5 (Y211) grain formations along the top of the $BaCeO_3$ reaction layer rather than Y_2O_3 particles. While Y_2O_3 can nucleate homogeneously in YBCO and is more frequently observed in YBCO films despite being less thermodynamically stable than Y211,[22-23] the high energy interfacial sites associated with the $BaCeO_3$ formation allowed for easier Y211 nucleation.

Microstructural Changes with Film Thickness for RABiTS2

Immediately, it can be seen by comparison between the 1.0 μm YBCO film on the RABiTS2 material (Fig. 3(a)) to that of the films on RABiTS1 (Fig. 1), that a significant improvement in film quality has occurred. The 1.0 μm film on the new substrates showed completed c-axis orientation, with no misoriented grains or porosity observed within the examined sample. Only contrast from stacking faults and the $c_\perp$ column boundaries are visible within the composite image.

The YBCO films greater than 1.0 μm on RABiTS2, were found to be similar to one another. The thickest of the films studied and the most remarkable of the microstructures, was the 6.4 μm thick YBCO film (Fig. 3(b)). The c-axis columns were clearly seen as growing the full film thickness, with numerous anti-phase boundaries observed which originate and terminate within the thickness of the YBCO layers. The mismatch across these boundaries was considered negligible through electron micro-diffraction.

The thicker YBCO samples also showed tubular or elongated porosity within the film, not observed in the 1.0 μm thick film. The triangular shape of the porosity imaged in Fig. 3(b), is the result of the cross-section having been cut at an angle relative to the normal of the film. Porosity was observed at all depths in the thick YBCO layers, and did not appear to be either connected to each other or long enough to reach the free surface of the YBCO film.

Further revealed under higher magnification was the presence of secondary phase particles within the YBCO layer, observable in the dark-field image of Fig. 3(c). The particles identified as CuO and Y_2O_3 through electron diffraction, formed as the result of the coalescence of excess Y and Cu oxide along stacking faults. The monoclinic CuO particles (Fig. 3(d)) were indexed as having a CuO [010] // YBCO [100] orientation relationship[18] were much larger in size than the Y_2O_3 particles (Fig. 3(e)) and grew as long continuous particles. The Y_2O_3 particles which held the Type A,[25-26] Y_2O_3 [110] // YBCO [100] relationship, were smaller and discontinuous in the film, typically forming separate from other Y_2O_3 or CuO particles. Staking faults were no longer observed in the thicker films as they had been in the 1.0 μm film. Volume changes associated with the formation of CuO and Y_2O_3 particles are believed to be responsible for the creation of the porosity observed within the thicker RABiTS2 films.

Reaction between the YBCO and CeO_2 layers within the RABiTS2 samples was also limited (Fig. 2(c)). Formations of both $BaCeO_3$ and Y211 were identified

Fabrication of High Temperature Superconducto

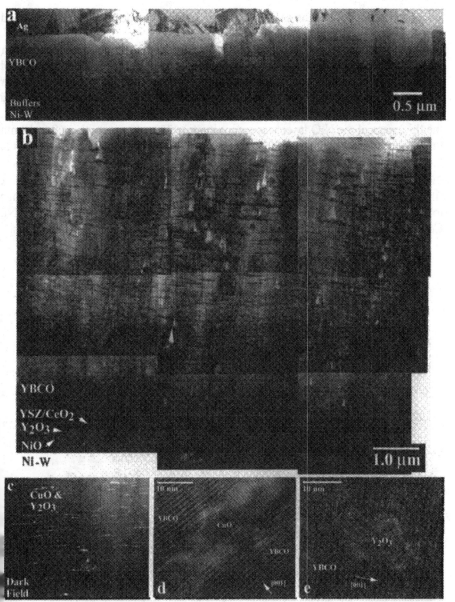

Fig. 3. Cross-section images of (a) 1.0 μm and (b) 6.4 μm thick YBCO films on RABiTS2. (c) Dark-field image of 6.4 μm thick film, and high-resolution images showing (d) CuO and (e) Y_2O_3 particles.

at the boundary of the 6.4 μm thick film, but the CeO_2 layer was found still intact. By comparison, the 6.4 μm thick YBCO film on the newer RABiTS2 structure showed much less interface reaction than that of the 3.0 μm thick film of the older substrates, even though the sample had seen twice the deposition time at elevated temperatures.

At the boundary between the Ni-W and Y_2O_3 seed layer, a NiO reaction layer was formed. While this also occurred for the RABiTS1 materials, it was found that the growth of NiO was limited within the Ni-W substrates. In the NiO layer formed within the RABiTS2 materials, a tungsten oxide layer was identified through high resolution EDS line scans (Fig. 4). For all the RABiTS2 samples, the thickness of the NiO layer was approximately 35 to 40 nm, with an 8 to 10 nm thick W-oxide layer and was fully developed or continuous in the NiO within the 6.4 µm sample. Where no W-oxide layer was found in the RABiTS1 materials, the NiO layer grew from 10 nm to over 150 nm for the 0.19 and 3.0 µm thick films, respectively. The tungsten oxide layer may have prevented further damage to the upper layers of the conductor stack by preventing the diffusion of Ni. Contributing to the enhanced performance of the YBCO on the newer substrates.

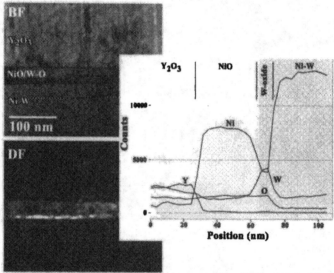

Fig. 4. Bright-(BF) and dark-(DF) field images of the NiO reaction layer between the Y_2O_3 buffer and Ni-W substrate showing the tungsten-oxide layer formed. An EDS nano-probe line scan across the interfaces shown in the images, illustrating the tungsten-oxide layer.

CONCLUSIONS

In combining microstructural investigation of the YBCO films with the x-ray characterization data, the leading cause of J_c loss for the RABiTS1 films was the degradation of cube texture. The formation of randomly oriented grains near the top of the YBCO film began around a thickness of 0.45 µm under the given PLD conditions for the RABiTS1 architecture. The amount of non c-axis grain formations increased with film thickness, along with the formation of a-axis

Fabrication of High Temperature Superconductc

oriented grains beginning with the 1.7 μm thick film. While the random grain formations dominated the upper portion of the film, they did not create a dead layer preventing c-axis growth up to the surface of the YBCO film.

The development of secondary phases along the YBCO/CeO$_2$ interface also played a significant role in the reduction of YBCO film properties for the RABiTS1 films. Transforming up to 0.1 μm of the YBCO film near the buffer surface into a non-conducting phase within the 3.0 μm thick film. This was not the case for RABiTS2 materials, which showed little reaction at the YBCO/CeO$_2$ interface.

The RABiTS2 samples showed only a modest decrease in J$_c$ with film thickness. In the thicker YBCO films on RABiTS2, micro-porosity associated with the formation of Y$_2$O$_3$ and CuO was observed, though its effects on electrical properties of the film are believed to be minimal. In addition, the formation of a tungsten oxide layer within the NiO reaction layer of the RABiTS2 samples appeared to have acted as a Ni diffusion barrier, further protecting the YBCO layer.

ACKNOWLEDGEMENT

Research sponsored by the United States Department of Energy, Office of Energy Efficiency and Renewable Energy, Office of Distributed Energy and Electric Reliability- Superconductivity Program. This research was performed at the Oak Ridge National Laboratory, managed by UT-Battelle, LLC for the United States Department of Energy under contract No. DE-AC05-00OR22725. The authors (K.J. Leonard, B.W. Kang, and S. Kang) are grateful to the Oak Ridge Associated Universities for their support.

REFERENCES

[1] S.R. Foltyn, Q.X. Jia, P.N. Arendt, L. Kinder, Y. Fan and J.F. Smith, *Appl. Phys. Lett.*, **75** [23] 3692-3694 (1999).

[2] S. Sievers, F. Mattheis, H.U. Krebs and H.C. Freyhardt, *J. Appl. Phys.*, **78** [9] 5545-5548 (1995).

[3] X.F. Zhang, H.H. Kung, S.R. Foltyn, Q.X. Jia, E.J. Peterson and D.E. Peterson, *J. Mater. Res.*, **14** [4] 1204-1211 (1999).

[4] J.H. Park and S.Y. Lee, *Physica C*, **314**, 112-116 (1999).

[5] T.G. Holesinger, S.R. Foltyn, P.N. Arendt, Q.X. Jia, P.C. Dowden, R.F. DePaula and J.R. Groves, *IEEE Trans. On Appl. Superconductivity*, **11** [1] 3359-3364 (2001).

[6] K.D. Develos, H. Yamasaki, A. Sawa and Y. Nakagawa, *Physica C*, **361**, 121-129 (2001).

[7] K.D. Develos, H. Yamasaki, A. Sawa, Y. Nakagawa, S. Oshima, and M. Mukaida, *Physica C*, **357**, 1353-1357 (2001).

[8] T.G. Holesinger, S.R. Foltyn, P.N. Arendt, H. Kung, Q.X. Jia, R.M. Dickerson, P.C. Dowden, R.F. DePaula, J.R. Groves and J.Y. Coulter, *J. Mater. Res.*, **15** [5] 1110-1119 (2000).

[9]K.J. Leonard, B.W. Kang, A. Goyal, D.M. Kroeger, J.W. Jones, S. Kang, N. Rutter, M. Paranthaman, and D.F. Lee, *J. Mater. Res.*, **18** [5] 1109-1122 (2003).

[10]O. Eibl and B. Roas, *J. Mater. Res.*, **15** [11] 2620-2632 (1990).

[11]S.R. Foltyn, E.J. Peterson, J.Y. Coulter, P.N. Arendt, Q.X. Jia, P.C. Dowden, M.P. Maley, X.D. Wu and D.E. Peterson, *J. Mater. Res.*, **12** [11] 2941-2946 (1997).

[12]H.Y. Zhai, I. Rusakova, R. Fairhurst and W.K. Chu, Phil. Mag. Lett., **81** [10] 683-690 (2001).

[13]A. Goyal, R. Feenstra, M. Paranthaman, J.R. Thompson, B.Y. Kang, C. Cantoni, D.F. Lee, P.M. Martin, E. Lara-Curzio, C. Stevens, D.M. Kroeger, M. Kowalewski, E.D. Specht, T. Aytug, S. Sathyamurthy, R.K. Williams and R.E. Ericson, *Physica C*, **382**, 251-262 (2002).

[14]A. Goyal, D.P. Norton, J.D. Budai, M. Paranthaman, E.D. Specht, D.M. Kroeger, D.K. Christen, Q. He, B. Saffian, F.A. List, D.F. Lee, P.M. Martin, C.E. Klabunde, E.C. Hatfield, and V.K. Sikka, *Appl. Phys. Let.*, **69** [12] 1795-1797 (1996).

[15]J.D. Budai, D.K. Christen, A. Goyal, Q. He, D.M. Kroeger, D. F. Lee, F.A. List, D.P. Norton, M. Paranthaman, B.C. Sales and E.D. Specht, U.S. Patent 5 968 877 (1999).

[16]S. Kang, A. Goyal, N.A. Rutter, K.J. Leonard, M. Paranthaman, S. Sathyamurthy, and D.M. Kroeger, *J. Mater. Res.*, in review (2003).

[17]B.W. Kang, A. Goyal, D.F. Lee, J.E. Mathis, E.D. Specht, P.M. Martin, D.M. Kroeger, M. Paranthaman, and S. Sathyamurthy, *J. Mater. Res.*, **17** [7] 1750-1757 (2002).

[18]K.J. Leonard, S. Kang, A. Goyal, K.A. Yarborough and D.M. Kroeger, *J. Mater. Res.*, in review (2003).

[19]J.G. Wen, C. Traeholt and H.W. Zandbergen, *Physica C*, **205**, 354-362 (1993).

[20]S. Bals, G. Van Tendeloo, G. Rijnders, D.H.A. Blank, V. Leca and M. Salluzzo, *Physica C*,**372**, 711-714 (2002).

[21]D.M. Hwang, T. Venkatesan, C.C. Chang, L. Nazar, X.D. Wu, A. Inam and M.S. Hegde, *Appl. Phys. Lett.*, **54** [17] 1702-1704 (1989).

[22]J.F. Hamet, B. Mercey, M. Hervieu, G. Poullain and B. Raveau, *Physica C*, **198**, 293-302 (1992).

[23]U. Scotti di Uccio, F. Miletto Granozio, A. Di Chiara, F. Tafuri, O.I Lebedev, K. Verbist and G. van Tendeloo, *Physica C*, **321**, 162-176 (1999).

[24]P.R. Broussard. L.H. Allen, V.C. Cestone and S.A. Wolf, *J. Appl. Phys.*, **74** [1] 446 (1993).

[25]K. Verbist, A.L. Vasiliev and G. Van Tendeloo, *Appl. Phys. Lett.*, **66** [11 1424-1426 (1995).

[26]A. Catana, R.F. Broom, J.G. Bednorz, J. Mannhart and D.G. Schlom, *Appl. Phys. Lett.*, **60** [8] 1016-1018 (1992).

THICKNESS DEPENDENCE OF J$_C$S IN YBCO, TL-2212 AND HG-1212 THICK FILMS

J.Z. Wu, R. Emergo and X. Wang

Department of Physics and Astronomy, University of Kansas, Lawrence, KS 66045

Abstract

Epitaxial growth of thick (typically a few microns) high-T$_c$ superconducting (HTS) films on metal substrates is desired for coated conductor applications in order to carry high electrical current of several thousands amperes per centimeter width. Serious degradation of the critical current (J$_c$) at large film thickness in YBa$_2$Cu$_3$O$_7$ (YBCO) films has raised a question on the mechanism related. To understand such mechanisms, we have investigated the thickness dependence of J$_c$s in several different types of HTS thick films including YBCO, Tl$_2$Ba$_2$CaCu$_2$O$_{7+\delta}$ (Tl-2212), and HgBa$_2$CaCu$_2$O$_{6+\delta}$ (Hg-1212). By comparing materials with different anisotropy that may affect the thickness dependence of J$_c$s, and microstructures that are resulted from different processing, we intend to pinpoint the mechanisms that are responsible for the degradation of the J$_c$ in HTS thick films.

1. Introduction

The second-generation high-temperature superconducting (HTS) wires employ the "coated conductor" approach, where long lengths of highly grain-oriented HTS coatings are epitaxially deposited on metal tapes [1,2]. This approach is key to overcoming the problem of "weak-link" current transport across high-angle grain boundaries (GBs) in HTS films and may offer an attractive alternative to the first-generation Bi-HTSs powder-in-tube tapes. The ultimate requirement for coated conductors is to fabricate tapes in kilometers which can carry large currents in the order of a few hundreds to thousands Ampere per centimeter width. Since typical YBCO thin films with their thickness less than 0.3 μm can carry J$_c$s of 4-5 MA/cm^2 at 77 K and zero applied magnetic field, the capability to carry critical current of a few hundreds to thousands Ampere per centimeter width will require YBCO films with thickness in the order of a few micrometers or thicker. Unfortunately, superconducting critical current density (J$_c$) drops drastically in YBa$_2$Cu$_3$O$_{7-\delta}$ (YBCO) thick films at large thickness. Several groups reported monotonic decrease of J$_c$s with increasing film thickness whether on single crystals or on metal substrates [3,4]. It is generally true that J_cs drop by a factor of 3-5 from several MA/cm^2 to about 1 MA/cm^2 when the film thickness increases from ~0.2 μm to ~1 μm. At larger thickness, J_cs continue to decrease but at a much slower pace. This pace, however, is different for YBCO films grown on single crystals and coated conductors. A much larger drop in J_c has been observed on coated conductors.

An interesting question arises on the mechanisms of the observed J_c vs. thickness behaviors: whether they are intrinsic to HTS materials due to, for example anisotropy of their crystalline and electronic structures (see Fig. 1), or extrinsic due to microsctructures formed via different processing. To answer this question, we have made a comparative study on the thickness dependence of J_c in $YBa_2Cu_3O_7$ (YBCO), $Tl_2Ba_2CaCu_2O_{7+\delta}$ (Tl-2212), $HgBa_2CaCu_2O_{6+\delta}$ (Hg-1212) films. In this paper, we describe our experimental results.

2. Sample preparation

Pulsed laser deposition (PLD) was employed to fabricate YBCO thick films on $SrTiO_3$ (STO) substrates. The (100) axis of the substrate is aligned with one of the edges of the substrates. A Lambda Physik KrF excimer laser was used for PLD with λ=248 nm and pulse duration of 25 ns. The laser energy density was estimated to be 1-3 J/cm^2 and the deposition rate was 0.06 nm/pulse at 5 Hz repetition rate. The substrates were silver pasted to the heater and deposition was made at 810 °C in 200 millitorr oxygen partial pressure. After the film deposition, the samples were annealed in situ at 520 °C and 350 torr oxygen partial pressure. Most films of the same thickness on STO of different vicinal angles were made in the same run by placing the substrates very close to each other to ensure a uniform deposition condition.

Tl-2212 films were made in crucible process in two different ways. $LaAlO_3$ (LAO) and MgO substrates were used. The thin Tl-2212 films with a thickness of 0.25—0.30 μm were prepared using the standard crucible technique [5]. The thicker Tl-2212 films with a thickness of 2.5—3.0 μm were made in a Tl-loss process [6]. Some of these Tl-2212 films were then used as the precursor films for Hg-1212 films and the conversion from Tl-2212 to Hg-1212 was made through Hg-vapor annealing them to replace Tl cations with Hg cations via cation exchange [7]. The details of the cation exchange process can found in [8,9]. Briefly, the Tl-2212 precursor films were torch sealed in an evacuated quartz tube together with an $HgBa_2Ca_2Cu_3O_x$ pellet as Hg vapor source and a $Ba_2Ca_2Cu_3O$ pellet as Hg absorber. The whole sample assembly was annealed in a furnace at 780°C for 4-9 hours. In order to optimize the oxygen content in these Hg-1212 films, post annealing in flowing oxygen at 300°C for 4 hours was applied to most

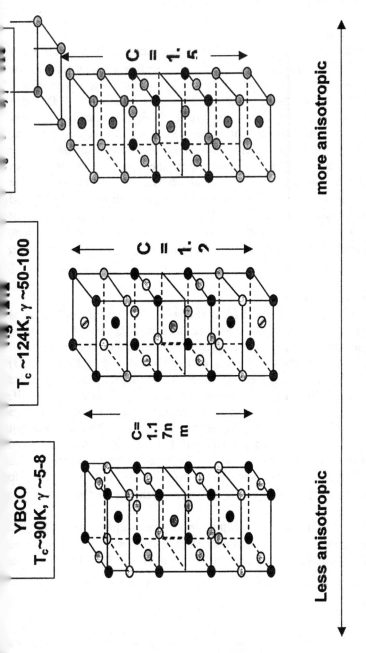

YBCO
$T_c \sim 90K$, $\gamma \sim 5-8$

$T_c \sim 124K$, $\gamma \sim 50-100$

C=
1.1
7n
m

C = 1.2

C = 1.5

Less anisotropic

more anisotropic

Fig. 2 Schematics showing variation of lattice structures and anisotropy in YBCO, Hg-1212 and Tl-2212.

Hg-1212 films. It should be mentioned the c-axis lattice constant for Tl-2212 is 1.48 nm, about 14% larger than that of Hg-1212. The conversion from Tl-2212 to Hg-1212 resulted in 14% thickness reduction in the c-axis as confirmed earlier in x-ray diffraction (XRD) θ-2θ scans and thickness measurements on Hg-1212 and their Tl-2212 precursor films [7]. The thickness of Hg-1212 thin and thick films used was, respectively, in the range of 0.21—0.26 μm and 2.1—2.6 μm.

3. Experimental results

The structures and phase purity of the films were characterized using x-ray diffraction (XRD) θ-2θ scans and pole figures on a Bruker x-ray diffractometer with four-circle goniometer. . The surface morphology of the films was examined using a LEO 1550 field-emission scanning electron microscope (SEM) and a Digital Instrument atomic force microscope (AFM). Both magnetic and electrical transport techniques were applied for investigation of the superconducting properties. The magnetic measurement was carried out on a Quantum Design SQUIDs magnetometer (MPMS7). For transport measurement, micro-bridges were defined on the films in a standard photolithography process. The bridge width was 20 μm or 40 μm and the length was 3 mm. Both resistivity (ρ) and critical current density (J_c) were measured on these micro-bridges at different temperatures (T) and applied magnetic fields (H). On the thicker Hg-1212 films, ion milling was applied to investigate the uniformity of ρ and J_c across the film thickness.

3.1. Structure and phase purity

XRD θ-2θ spectra were taken in most films studied in this experiment and all of them showed that the films have the c-axis orientation. XRD pole figures were acquired on some films to confirm the epitaxy. For example, (103) poles were acquired on Hg-1212 films and (105) poles, on the Tl-2212 films. The four (103) poles [or (105) poles for Tl-2212 precursor films] are 90 degree apart in φ angles and aligned with the (100) [or (010)] axes of the substrate. The full-width-half-maximum (FWHM) of the (103) poles for thinner Hg-1212 films was in the range of 0.5-0.8 degree while that for the thicker Hg-1212 films, 0.6-1.1 degree. It should be noticed that these FWHM's in Hg-1212 films are comparable to that of their Tl-2212 films. This suggests that the original epitaxial structure was transferred from the Tl-2212 precursor films to Hg-1212 films of thickness up to 2.6 μm and the texture degradation during the cation exchange was minimal [9].

The phase purity of these HTS films can also be estimated from the XRD θ-2θ spectra. The amount of impurity phases on all these three types of films was negligibly small. However, certain amount of unconverted Tl-2212 phase was

Fabrication of High Temperature Superconduct(

visible in the XRD θ-2θ spectra for Hg-1212 thick films (thickness > 0.5 μm). The volume portion of the Hg-1212 can be estimated from *Hg-1212 left in the film (%)= $I_{Hg}/(I_{Hg}+I_{Tl})$*. Here I_{Hg} and I_{Tl} are respectively the intensity of the strongest peak for Hg-1212 [(005) peak] and Tl-2212 [(0012) peak] in their XRD θ-2θ spectra. For thin films, the volume portion of Hg-1212 above 90% can be routinely obtained and the highest so far observed is above 99%. If processed at more or less the same conditions, more remaining Tl-2212 phase was observed in thicker films [10]. This suggests that much longer processing time may be necessary to convert thicker Tl-2212 films to Hg-1212 films. To confirm this, we processed 3.0 μm thick Tl-2212 films at 780 °C for different periods in cation exchange process. It was found that the volume portion of the Hg-1212 film increases monotonically with the annealing time and exceeds ~92% when the Hg-vapor annealing time is 12 hours. Although the Hg-1212 volume portion still increases linearly with the cation exchange time up to 12 hours, longer processing time up to 24 hours did not yield the complete conversion. The mechanism remains unclear. A possibility is that the Hg-vapor pressure inside the quartz ampoule may begin to decrease after certain period, reducing the efficiency of the cation exchange afterwards. In any cases, the Hg-1212 films used in this experiments all have the volume portion of Tl-2212 less than 10%.

3.2. Surface morphology

Fig. 2 compares surface morphology of YBCO, Tl-2212 and Hg-1212 films at three different thickness: thin film regime around 0.2 μm, medium one of 1.3-1.5 μm, and thick one of 2.6-3.0 μm. All three thin films have smooth surface morphology with either particulates (for YBCO) or secondary phases-Tl-rich particles (on Tl-2212 and Hg-1212) visible on the surface of the film. Notice the scale for the YBCO films is slightly different from that for Tl-2212 and Hg-1212 films. With increasing thickness, for example at 1.5 and 3.0 μm thickness, the morphology of the films experienced considerable changes as shown in Fig. 2. On YBCO films, the surface became much rougher and the size of the particulates increased significantly. On Tl-2212 films, on the other hand, uniform porous structures were seen in the film thickness range from 0.6 to 3.0 μm. Some of the 3.0 μm thick Tl-2212 films were thinned using ion milling and SEM pictures were acquired at different depths [11]. Similar morphology was confirmed at different depths, suggesting a uniform structure across the thickness of the Tl-2212 thick films. The surface of the Hg-1212 films is dense but rougher with many voids of dimensions in the range of sub-μm to several μm's. Such voids have been observed on most Hg-1212 films converted from Tl-2212 films, despite the thickness and processing conditions employed [9]. Since no voids were observed on Hg-1212 films converted from Tl-1212 precursor films, we suspect

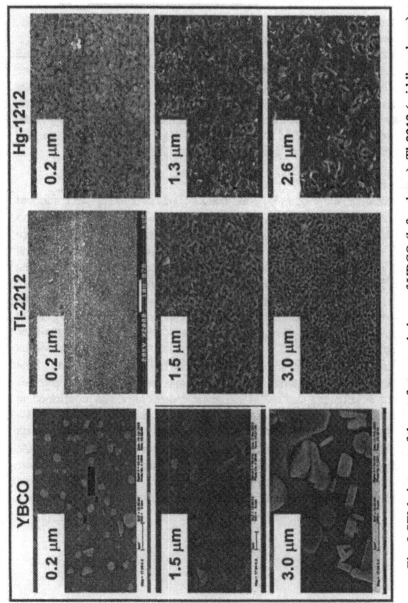

Fig. 2 SEM pictures of the surface morphology of YBCO (left column); Tl-2212 (middle column); and Hg-1212 (right column) films of a 0.2 μm, 1.3-1.5 μm, 2.6-3.0 μm thickness. Notice the scale for YBCO films is 2.5 times smaller than that for Tl-2212 and Hg-1212 films.

Fabrication of High Temperature Superconduct

that the extra Tl cations in Tl-2212, by making a way out, are the major reason for the formation of the voids.

3.3. T_c and J_c (T,H)

The T_cs of the HTS films were measured using a Quantum Design SQUID magnetometer as well as using the standard four-probe electrical transport technique. The results of the magnetic and transport measurements were consistent. Fig. 3 depicts the zero-field-cooled magnetic moment (M) of three YBCO films [Fig. 3(a)] and three each of Tl-2212 and Hg-1212 films [Fig. 3(b)] as a function of temperature (T) in a 10 Oe magnetic field (H) applied along the c-axis of the film. The three samples for each material had different thickness: thin film regime around 0.2-0.6 μm, medium one of 1.3-1.5 μm, and thick one of 2.6-3.0 μm. The T_c's of three YBCO films were about 87-91 K. The T_cs of the Tl-2212 films were 100-102 K, and that for Hg-1212 films, 120-122 K, about 20 K higher than that of the original Tl-2212 precursor films. Overall, sharp transition near Tc was observed on most films studied in this experiment. The ρ measurement made on some films confirmed linear ρ vs. T curves above T_c's. ρ at room temperature was typically less than 300 μΩ cm for thin films while it was 20-50% higher in thick films.

The J_c of the Hg-1212 film can be estimated from the M-H hysteresis loops using the Bean model. Fig. 4 plots J_c vs. thickness curves for YBCO, Tl-2212 and Hg-1212 films at 77 K and self field. It should be mentioned that for each thickness, many films were made and characterized and the J_cs in Fig. 4 are the representative J_cs of samples showing consistent behaviors. As shown in Fig. 4(a), J_c's of the three different types of films experienced a similar decrease with increasing thickness. This J_c decrease is much bigger initially when the film thickness is smaller or near 1μm. For example, at 77 K, the J_c of the 0.25μm thick Hg-1212 films is around 3.8 MA/cm^2 and it reduces by a factor of four when the film thickness is increased to 1.3 μm. When the film thickness is further increased to about 2.6 μm, the J_c remains nearly the same. This J_c vs. thickness behavior demonstrated in Fig. 4(a) resembles what has been reported before on YBCO films [3,4]. This is not surprising since the same PLD technique was employed. The reduction of J_c with thickness is more or less uniform at different temperatures. For example, at 100 K, the J_c of 1.39 MA/cm^2 for the 0.25 μm thick Hg-1212 film reduces to 0.41 MA/cm^2 for the 2.6 μm thick films--a factor of 3.4, while at 77 K and 5 K, this ratio is 4.1 and 3.3, respectively. Fig. 4(b) depicts the same J_c vs. thickness curves for YBCO, Tl-2212 and Hg-1212 films on the normalized scale. It is clearly shown that Tl-2212 films have the smallest thickness dependence while Hg-1212, the largest. This seems to suggest the thickness dependence may not have a direct correlation with the anisotropy of the

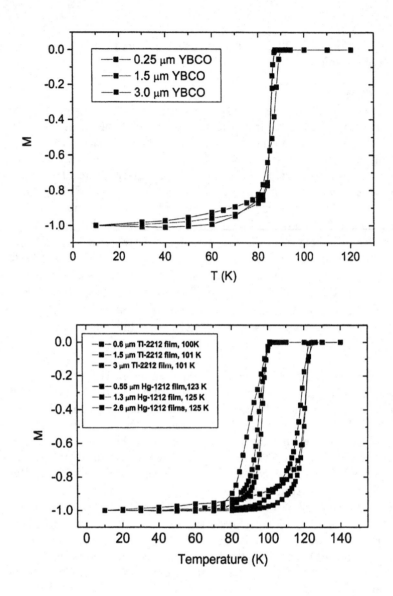

Fig. 3 Zero-field-cooled M-T curves of (a) three YBCO films of 0.25, 1.5 and 3.0 μm thickness; and (b) three Tl-2212 films of 0.6, 1.5 and 3.0 μm thickness and the three Hg-1212 films converted from these Tl-2212 films. H=10 Oe.

Fabrication of High Temperature Superconduct

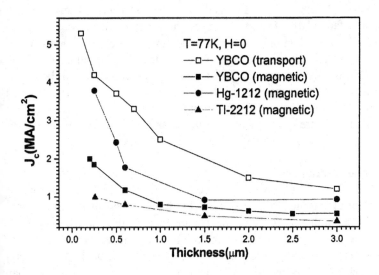

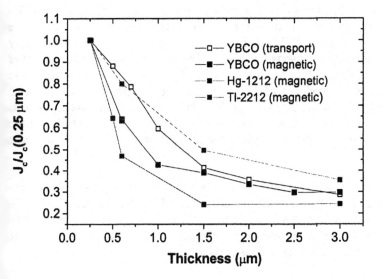

Fig. 4 A comparison of thickness dependence of J_cs of YBCO, Tl-2212 and Hg-1212 films in (a) original and (b) normalized scale at 77 K and self field.

material. However, our recent study on the depth profile of J_cs in 2.6-3.0 μm thick Tl-2212 and Hg-1212 films revealed considerable degradation of films near the substrate/film interface [10]. The thickness of the degraded layer was over 1.0 μm and it reduced to less than 0.5 μm for films of 1.5 μm thickness. On the other hand, the near interface film degradation was confirmed minimal in our PLD YBCO films. Taking this account, the J_c for Tl-2212 and Hg-1212 will be raised at large thickness, which means that the J_c vs. thickness curve for Tl-2212 and Hg-1212 in Fig. 4 will be flatter if degradation problem is solved. Although it is difficult to visualize where the J_c-thickness curve for Hg-1212 will be, it is clear that the curve for Tl-2212 is much flatter than that for YBCO.

4. Discussions and summary

There are two distinctive differences between YBCO and Tl-2212. One is the anisotropy in their electronic structures. YBCO is more or less three-dimensional with the anisotropy index γ<10 (γ is the square-root of the ratio between the effective mass along c-axis and ab-plane). Tl-2212, on the other hand, is extremely anisotropy and has been considered two-dimensional system with γ>500. Consequently, the penetration depth of the Tl-2212 is much larger than that of YBCO. If J_cs are only carried in the surface layer of thickness on the order of penetration depth, they should have much less thickness dependence in Tl-2212. Another factor that may affect the thickness dependence of J_c is the microstructure. Indeed, YBCO and Tl-2212 thick films have distinctively different microstructures as shown in Fig. 2. The former is dense while the latter is porous. A simple minded argument in favor of porous structure having smaller or none thickness dependence of J_c is the dramatically increased surface area, allowing penetration of the magnetic field through the whole film thickness. This is supported by the less thickness dependence of J_c in porous YBCO films made in solution process [12]. Further experiments are certainly needed to clarify these mechanisms.

In summary, we have fabricated thick films of YBCO, Tl-2212 and Hg-1212 and investigated the thickness dependence of J_c in these films at 77 K and self field. Our results suggest a lesser thickness dependence of Jc may be obtained in HTS films with higher anisotropy. On the other hand, porous microstructure may be favorable for higher J_cs in thicker HTS films.

Acknowledgement

JZW acknowledges supports from NSF and DOE for this work. RE and XW were supported by AFOSR.

REFERENCES

1. D.K. Finnemore, K.E. Gray, M.P. Maley, D.O. Welch, D.K. Christen, and D.M. Kroeger, Physica C 320, 1 (1999).
2. M. Paranthaman, A. Goyal, R. Feenstra, T. Izumi and V. Selvamanickam, "2002 MRS Workshop on Processing and Applications of Superconductors", Gatlinburg, TN, July 31-Aug. 2, 2002.
3. X.D. Wu, S.R. Foltyn, P.N. Arendt, W. R. Blumenthal, I. H. Campbell, J. D. Cotton, J. Y. Coulter, W. L. Hults, M. P. Maley, H. F. Safar, J. L. Smith, Appl. Phys. Lett. 67, 2397 (1995).
4. S.R. Foltyn, Q.X. Jia, P.N. Arendt, L. Kinder, Y. Fan, and J.F. Smith, Appl. Phys. Lett. 75, 3692 (1999).
5. M. P. Siegal, E. L. Venturini, B. Morosion, and T. L. Aselage, J. Mater. Res. 12, 2825 (1997).
6. B. Hammond, G.V. Negrete, L.C. Bourne, D.D. Strother, A.H. Cardona, and M.M. Eddy, App Phys. Lett. 57, 825 (1990).
7. J.Z. Wu, S.L. Yan, and Y.Y. Xie, Appl. Phys. Lett. 74, 1469 (1999).
8. S.L. Yan, Y.Y. Xie, J.Z. Wu, T. Aytug, A.A. Gapud, B.W. Kang, L. Fang, M. He, S.C. Tidrow, K.W. Kirchner, J.R. Liu and W.K. Chu, Appl. Phys. Lett. 73, 2989(1998).
9. S.L. Yan, J.Z. Wu, L. Fang, Y.Y. Xie, T. Aytug, A.A. Gapud, and B.W. Kang, J. Appl. Phys 93, 1666 (2003).
10. Z.W. Xing, Y.Y. Xie, J.Z. Wu and A. Cardona, to appear in Physica C.
11. X. Wang, R. Emergo, and J.Z. Wu, unpublished.
12. R. Finnstra, M. Feldmann, T. Holesinger, DOE superconductivity program peer review, Washington DC, July 2003.

Fabrication of High Temperature Superconduct▪

DAMAGE EFFECT IN HTS IRRADIATED BY U FISSION FRAGMENTS

Alberto Gandini,
Physics Department and TCSAM
University of Houston
Houston, TX, 77204-5005, USA

Roy Weinstein,
Physics Department and TCSAM
University of Houston
Houston, TX, 77204-5005, USA

ABSTRACT

We discuss the effect that the volume of irradiation-induced damages has on the transport current properties of superconductors. We argue that irradiation-damages do not only act as pinning centers, hence increase J_c, but do also reduce the area through which current percolate, hence reduce I_c. The reduction of the percolating current area is discussed within the framework of percolation theory. Among the several irradiation processes, the U/n process is a leading method to increase I_c in HTS. In the U/n process, damages are produced by uranium fission fragments. The induced defects in this case are particular, as they change in size and morphology along the fission fragment range. Then, in here, we particularly focus on size and morphology of the U/n fission fragment defects, and on their effect on the percolating current area.

BACKGRAOUND

To achieve high critical current in high temperature superconductors (HTS) pinning centers must be introduced within the HTS matrix [1]. Pinning centers are not-superconducting regions. Theoretical works as well as experimental data indicate that the best pinning centers have dimensions of the order of the superconducting coherence length [1 - 2]. Pinning centers may be introduced by chemical doping [3 - 4], and by irradiation [5 - 6]. Irradiation-induced damages have been proven to be the most effective types of pinning center; however, their full potential has not been explored. Indeed, most of the studies in the last decade have been focused only on the low pinning density regime, where critical current, $_c$, is observed to rapidly increase with the pinning-center density. Studies are discontinued at pinning-center density beyond which I_c begins to fall-off. Little attention has been paid to this effect [7 – 8] (i.e., the I_c fall-off at large density.)

The I_c fall-off is usually attributed mostly to the reduction of T_c, and the fall-off has been considered as an inevitable secondary effect [5]. For this reason, most of the studies have been carried at low radiation fluence, below the peak of I_c attainable.

In recent works [7 - 8], the mechanisms of the I_c fall-off at large fluence have been addressed. The driving idea is that in order to full explore the potential of irradiation-induced pinning centers the mechanisms, which limit the I_c enhancement, should be fully understood. As the mechanisms are identified and their effects quantified, a way to reduce them may be found as well. The main result of these works was to show that defects "collectively" result in a reduction of the percolating current area (this is also called the flow-area.)

The reduction of the flow-area, as the defect density increases, is most naturally discussed within the framework of percolation theory. Percolation theory describes the effects on the conductivity of a conductor as holes are randomly inserted in the sample [9 - 10]. Typically, it is applied to normal metal; though a priori the theory shall apply also to superconductors [11]. As one may see, a similar scenario occurs in irradiated HTS. In fact, irradiation damages are randomly distributed in non-superconducting amorphous regions. Hence, from a current conduction point of view those damages behave as holes. It is as if the damaged regions were removed from the HTS matrix, and they were not contributing to the current transport.

Weinstein et al. [7] investigated the effect of the size of columnar pinning centers on the magnetic properties of HTS. In [7], it has been shown that the volume of columnar pinning centers limits the maximum pinned field, and the percolating current area is severely reduced at a point which current can no longer flow within the sample, even at a relative low columnar pinning center density.

Gandini et al. [8], using a phenomenological model that makes use of the reduction-of-flow-area idea, showed that the I_c enhancement is limited by the volume of the defects that collectively reduce the area through which current flows. The model was tested against data on irradiated Ag/Bi-2223 tapes, and it was shown to fit experimental data well. In particular, the model indicates an interesting feature: if the volume of a defect is reduced by a factor of two, then I_c becomes larger up to a factor of two may be achieved [12 - 7].

We begin to realize that the size of a pinning center is important not only because the pinning potential depends on it, but also because the volume of the defects collectively limits severely the maximum number of pinning centers that can be added, which limits the maximum critical current achievable.

Two independent experiments [13 - 14] also showed that irradiation is mostly effective when it creates amorphous damages within the HTS. As it is discussed in the next section, not all forms of radiation create amorphous regions, although all forms of radiation induce vacancy-type damage (in here we consider only protons, neutrons and heavy ions irradiation.) [13 - 14] show that the vacancy-type damages are much less effective as a pinning center than the amorphous defects.

However, just as the amorphous damages, so do the vacancy-type damages reduce the flow-area. A deeper understanding of the volume-effect of the irradiation-induced damage, in particular of the vacancy-type damages as they reduce the flow-area while only marginally increase I_c, is then of great interest in the process of increasing the critical current.

In the following, size and morphology of U/n fission fragment [15] is investigated, and its effect on I_c is discussed.

THE U/n PROCESS

The U/n process has been proven as one of the most effective process for increasing I_c [15 - 16]. In the U/n process, U is added to the HTS precursor powder. The doped powder is processed following the standard HTS preparation methods, then samples are irradiated with thermal neutrons. Thermal neutrons fission some of the U atoms. The product of the fission is two heavy ions; in the following we refer to those as fission fragments. Fission fragments kinetic energy is discussed in [15]. As discussed in the introduction, the volume of the damage has a large effects on the maximum achievable I_c. In this paper, we then discussed in detail the damage due to the U/n process fission fragment, as this process is one of the leading processes to increase I_c.

ENERGY LOSS AND DAMAGES BY HEAVY ION IN HTS

As a swift ion travels through matter, it transfers energy to the target atoms. The energy of the ion is transferred via two different mechanisms, which occurs simultaneously. Ion may lose energy via Coulomb interaction with the electron [17]. In this case, the energy loss per unit of length is commonly referred in the literature as S_e. Ions may also lose energy via Coulomb force with the target atom nuclei, the energy loss per unit of distance in this case is called S_n [17]. Typically, the ion and energies used to irradiate HTS are such that: $S_e \sim 1000$ larger then S_n.

As the ion travels through matter and its energy is transferred to the target atoms, damage in the target lattice may form. Size and density of defects vary with the energy-loss mechanisms [17 - 19].

Defect Production Via S_e

The size and the morphology of the S_e-induced defects vary significantly with S_e. For $S_e < 8$ KeV/nm, no defects are formed. Whereas for $S_e > 8$ KeV/nm amorphous regions begin to form, in YBCO as well as in BiSCCO. The physics is that the amount of energy transferred below 8 KeV/nm is not enough to locally melt the HTS, hence to form an amorphous region [17 - 18]. The diameter and the length of these amorphous regions vary with S_e. A collection of data of defect diameter versus S_e is presented in [7]. At low S_e, an ion track looks like a string of beads separated by gaps of undamaged regions. As S_e increases the gaps between the amorphous damages diminishes. At $S_e > 35$ KeV/nm, continuous amorphous columnar defects form. The diameter of the damage has been extensively studied, whereas the length of the damaged segment, relative to the ion range, has not; though some experiments have been done [18]. An experiment done with gold and silver ions showed that for $S_e \sim 35$ KeV/nm full columns are formed, while for $S_e \sim 25$ KeV/nm amorphous regions form only along $\sim 50\%$ of the ion range.

Defect Production Via S_n

Energy loss via S_n leads to the formation of defects provided the energy transferred to the nucleus, E_t, is larger than the target displacement energy, E_d. The target displacement energy is the energy that must be given to a target atom to remove it from its equilibrium position. E_d is typically of the order of 20-25 eV. For $E_t > E_d$ vacancies and clusters of vacancies may form. Typically, vacancies are too small to be observed by SEM [17], though cluster of vacancies have been observed to be ~ 3 nm in diameter [20]. The S_n-induced damage increases the normal state resistivity as well as it decreases the critical temperature T_c. The change of resistivity and of T_c has been shown to scale with the defect density, expressed in term of number of displacements per atom, *dpa*. For example, Hensel et al. [17] showed that for several ions, with various energies and different values of S_n, the change of resistivity is equal for equal *dpa* amount.

ENERGY LOSS AND DAMAGE OF FSSION FRAGMENT

S_e for fission fragment has been calculated using the TRIM program [21]. The result is shown in figure 1. It is seen that as the ion travel ~ 3 µm S_e falls below the 8 KeV/nm threshold to form amorphous regions. A collection of data from the literature in combination with the TRIM results shown in figure 1 indicate: (i) along the first ~ 3 µm, fission fragment tracks consist of broken column followed by string of beads; (ii) fission fragment tracks have an amorphous core of radius ~ 1.8 nm (at the beginning of the track).

As we are interested in the net volume of the defect, which is subtracted from the current transport area, we shall notice two important aspects of irradiation damage. i) The amorphous core is surrounded by a damaged region (due to the shock wave) twice as large as the amorphous core itself [18]. ii) Because superconductivity is only fully restored into the matrix away from a non-superconducting region over a distance of the order of the coherence length, ξ, the volume over which superconductivity is destroyed is larger than the actual geometric volume of the damage. Then the volume of the first part of a fission fragment track is $\sim \pi(r + \xi)^2 l \sim 0.54 \times 10^{-15}$ cm^3, where $r \sim 3.6$ nm is the radius of the damaged area (amorphous core + shock wave), $\xi \sim 4$ nm at 77 K in Bi-2223 [22], and $l \sim 3$ µm is the length of the amorphous part.

As a fission fragment passes the 3µm, although most of the energy is still lost via S_e (see figure 1), no amorphous regions are formed since S_e is below the 8 KeV/nm amorphization-threshold. However, after the first ~ 3 µm a fission fragment still transfers enough energy via S_n to the target nuclei to displace some of them. In fact, the transferred energy, E_t, to the target atoms is $\sim 11 \times S_n$, as on

Fabrication of High Temperature Superconduct

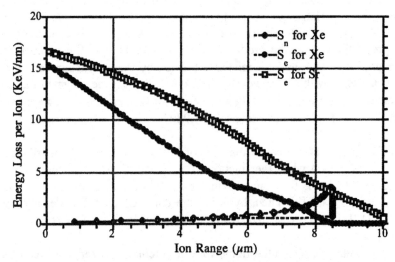

Figure 1. Energy loss per unit of length (KeV/nm) versus ion (i.e., fission fragment) range, for two examples of fission fragments. Fission fragment Xe has initial energy of 65 MeV. Fission fragment Sr has initial energy of 96 MeV. Close circles and open squares are energy loss via S_e for Xe and Sr, respectively. Open diamonds are energy loss via S_n for Xe; energy loss via S_n for Sr is similar to S_n for Xe, hence it is not shown in figure.

average fission fragments encounter 1 target atom every 1.1 nm. Then, it follows from Figure 1 that E_t is larger then E_d along most part of the fission fragment range, hence a fission fragment produces vacancy-type defect along most part of its range. TRIM [21] calculation shows that each fission fragment creates several vacancies along its path. The number of vacancies depends on the particular pair of fission fragments form during the U fission [15]. In fact, as U fissions, several different ions-pairs (e.g., fission fragment pair) may form, e.g., Xe and Sr. On average, the number of vacancies per fission fragment per cm, P, is $\sim 0.42 \times 10^8$ vac/(cmff) and $\sim 1.2 \times 10^8$ vac/(cmff) for Xe and Sr, respectively. The pair Xe-Sr is just one of the many possible U fission fragment pairs. The above averages are calculated along the second part of the fission fragment track, where S_e is too low to create damage.

The volume of damage along the second part of the fission fragment track is more complicated to evaluate, and the result has a larger error as the size of these damages is too small to be studied systematically with SEM. However, we think that a good estimate of the S_n-induced damage can be obtained by performing the following percolation theory-based analysis. Hensel et al. [17] showed that conductivity in YBCO as the ions lose energy via S_n (while the S_e is below the amorphous threshold) varies with dpa. In particular, they show that at low dpa, $\sigma \sim 1/(1400 \times dpa+1)$, whereas at high dpa decreases more rapidly, and $\sigma \sim 0$ at $dpa \sim 0.025$. Here, we use Hensel et al. [17] data and re-interpret them in term of percolation theory. Accordingly to the percolation theory, the conductivity decreases proportionally to the product: $V_{Sn} \times n$, where: 1) n is the number of defects per unit of volume, and it is related to the dpa by the simple expression: $n = (\#atoms/cm^3) \times dpa$, with $(\#atoms/cm^3) \sim 7.07 \times 10^{22}$; 2) V_{Sn} is the volume of a S_n-induced defect. Then, it follows that $V_{Sn} \times n = V_{Sn} \times (\#atoms/cm^3) \times dpa = 1400 \times dpa$. Hence, the volume of a S_n-induced defect is $\sim 1.98 \times 10^{-20}$ cm^3, which gives a radius of 1.7 nm.

Percolation theory also predicts that, in conductor with randomly distributed spherical defects, $\sigma \sim 0$ when the fraction of the total damaged volume, $V_c = V_{damage}/V_{tot}, \sim 0.03$ (this is the main striking result of such theory). The volume fraction is $\sim \exp(-V_{Sn} \times n)$ [9]. Then, by using the data at high defect density by Hensel et al. [17], i.e., $\sigma \sim 0$ at $dpa \sim 0.025$, we get that $V_{Sn} \sim 1.98 \times 10^{-21}$ cm^3, hence: $r \sim 0.8$ nm.

Our reinterpretation, within the framework of percolation theory, of Hensel's data [17] shows that S_n-induced damages may be treated as spherical defects with radius $\sim 0.8 - 1.7$ nm, the reduction of conductivity is due to the loss of percolation area.

Hensel et al. data [17], as well as other experiments in the early '90's, were done at $T > T_c$. To support our analysis we search the literature for similar data obtained at $T < T_c$. Weaver et al. [23] show an interesting behavior of J_c. It is shown that the ratio of J_c after irradiation to J_c before irradiation, $J_c(\Phi)/J_c(\Phi=0)$, decreases as $\sim \exp(-7.7 \times 10^{-17} \times \Phi)$, where Φ is the proton fluence. We calculate, using the TRIM program, the dpa per unit of proton fluence, then by carrying the same analysis as above, we obtain the radius of the proton damage to be ~ 1.8 nm . This is also in agreement with the estimate by Civale [20]. We should notice at this time that the proton used in Weaver et al. [23] experiment had $S_e \sim 4$ eV/nm, much lower than the threshold for amorphization, and $S_n \sim 2.7 \times 10^{-3}$ eV/nm.

In conclusion, Hensel et al. [17] and Weaver et al. [23] data, re-interpreted in terms of percolation theory, indicate that the change in σ, by ions which have S_e below the amorphization threshold but S_n large enough to create vacancy-type defects, is due to the reduction of percolation area. Furthermore, the damage produced via S_n is composed of spherical defects of radius $\sim 0.8 -1.8$ nm. We shall shortly comment on the magnitude of the radius found. Since S_n damages are vacancy-type damages, one should indeed expect them to be at most of the order of few unit cells (unit cell size ~ 1 nm), in agreement with the above found value of $\sim 1.5 -1.7$ nm. This is also in agreement with Civale [20] observation.

We have now enough information to estimate the volume damage along the second part of the track. We use an average radius for the S_n-induced vacancy-type damage of ~ 1.6 nm, and an average number of vacancies per fission fragment

, P, of $\sim 0.8\times10^8$ vacancies/cm. Then, $V_{Sn} = (4\pi/3)r^3 \times P \times l$, where $r \sim 1.6$ nm is the radius of a spherical S_n-induce defect, and $l \sim 6$ μm is the average length of the second half of the fission fragment track. Hence, $V_{Sn} \sim 0.82\times10^{-15}$ cm^3. Surprisingly, V_{Sn} is comparable to the amorphous damage along the first part of the track.

The total volume of a fission fragment defects, V_d, is given by: $V_d = V_{Se} + V_{Sn}$ $\sim 1.37\times10^{-15}$ cm^3. The effect of this damage is to reduce the flow-area by a factor of $\sim$ exp $(-1.37\times10^{-15} x\!f\!f)$. The combine volume of the damaged induced via S_e with the damage volume via S_e turn out to be on the same order as the volume empirically found $\sim$ exp $(- 4.1\times10^{-15} x\!f\!f)$.

SUMMARY

We have discussed the effects of damage volume on critical current, and estimated the value of the U/n fission fragment volume. We have shown that fission fragment track is composed of two regions. Along the first ~ 3 μm of the their range fission fragments form amorphous quasi-columnar defects of ~ 3.6 nm in radius; along the last ~ 6 μm of their range fission fragments instead produce vacancy-type spherical defects of radius ~ 1.6 nm. The density of the vacancy-type defects was calculated with the TRIM program to be $\sim 0.82\times10^8$ #vacancies/cm per fission fragment. We estimated that the volume of the damage along the first part of the track is comparable to the volume along the second part. Hence, both parts contribute in equal manner to decrease the current percolating area. However, as it has been discussed, the amorphous part of the fission fragment damage is an efficient pinning center, while the vacancy-type damage has a low pinning efficiency. In other word, the second part of the fission fragment while reduces the current percolating does not increase J_c. We finally argue that a reduction of the volume of second part of the track may lead to a smaller reduction of the current percolating area, hence it may lead to a larger I_c enhancement as now achievable [24]. The vacancy-type damage along the second part of the fission fragment range may be reduced by annealing [25]. A preliminary experiment has been done in which U/n processed Ag/Bi-2223 tapes have been annealed at 400 Celsius. The first results show that I_c increases after annealing by $\sim 20\%$.

ACKNOWLEDGMENT

This work was supported in part by the U.S. Army Research Office, the Welch Foundation, the State of Texas via the Texas Center for Superconductivity and Advance Material (TCSAM).

REFERENCES

[1] L. Civale, "Irradiation Processing for Flux Pinning Enhancement in High-T$_c$ superconductor", Processing and Properties of High-Tc Superconductors, Vol 1, Chap. 8, World Scientific.
[2] D. Nelson, V. Vinokur, Phys. Rev B 48 13,060 (1993).

[3]R. Weinstein and Ravi-Persad Sawh, "A Class of Chemical Pinning Centers Including Two Elements Forein to HTS", Physics C 383, 438-444 (2003).

[4]Y. Ren, R. Weinstein and Ravi-Persad Sawh, "New Chemical Pinning Ceneter from Uranium Compound in Melt-Textured YBCO", Physics C 282-287 (1997) 2275-2276.

[5]R. C. Budhani, "Studies of Vortex Localization Columnar Defects in High Temperature Superconductor", Studies of High Temperature Superconductor, Vol. 33, Ed. Anant Narlikar, 2000 Nova Science Publishers.

[6]L. Civale, "Vortex Pinning and Creep in High-Temperature Superconductors with Columnar defects", Supercod. Sci. Technol. 10, A11-A28 (1997).

[7]R. Weinstein, A. Gandini, Ravi-Persad Sawh, D. Parks, and B. Mayes, "Improved Pinning Regime by Energetic Ions Using Reduction of Pinning Potential", Physics C 387 (2003) 391-405.

[8]A. Gandini, R. Weinstein, D. Parks, R. P. Sawh, and S.X. Dou, "On the limiting Mechanism of Irradiation Enhancement of I_c", Presented at the 2002 Applied Superconductivity Conference (ASC 2002), Agust 4-9, 2002 Houston, Texas, USA. Accepted for publication on IEEE Transaction on Applied Superconductivity.

[9]D. Stauffer, and A. Aharony, Introduction to Percolation Theory 2nd edn (1992), (Bristol, Pa: Taylor & Francis).

[10]M.D. Rintoul, "Precise Determination of their Void Percolation Threshold for Two Distributions of Overlapping Spheres', Phys. Rev. E, (62) 68-725 (2000).

[11]B. Zeimetz, B.A. Glowacki, and J.E. Evetts, "Applivation of Percolation Theory to Current Transport in Superconductors", Eur. Phys. J. B 29, 359-367 (2002).

[12]A. Gandini, "Improvement of Critical Current, Irreversibility Field, and Anisotropy in BiSCCO Tape by the U/n Method", Ph.d. Dissertation, University of Houston (2001).

[13]R. Weinstein, Ravi-Persad Sawh, J. Liu, D. Parks, Y. Ren, V. Obot, C. Foster, "Threshold for Creation of Ionization Pinning Centers in YBCO by Heavy Ions", Physics C 357 (2001) 743-746.

[14]Ravi-Persad Sawh, R. Weinstein, Y. Ren, V. Obot, and H. Weber, "Uranium Fission Fragment Centers in Melt-Textured YBCO", Physics C 341-348, 2441-2442 (2000).

[15]R. Weinstein, Invited Paper, "An Overview of U/n Processing," Proc. of 12th International Symposium of Superconductivity (ISS-99) Morioka, Japan, Oct. 17-19, 1999, Advances in Superconductivity XII, Springer-Verlag, 521-526 (2000).

[16]S. Tönies, H. W. Weber, Y. C. Guo, Shi X. Dou, R-P. Sawh, R. Weinstein, "Improved In-field Behavior of Uranium Doped BiSCCO Tapes by Enhanced Flux Pinning," IEEE Transactions on Applied Superconductivity, Vol. 11 (March 2001), pg.3904-3907. (ASC-2000, Virginia Beach, Sept 17-22, 2000).

[17]R. Hensel, B. Roas, S. Henke, R. Hopfengartner, M. Lippert, J.P. Strobel, M. Vildic, and G. Saemann-Ischenko, "Ion Irradiation of Epitaxial $YBa_2Cu_3O_{7-x}$ Films: Effects of Electronic Energy Loss", Phys. Rev. B 48, 4135 (1990).

[18]Y. Zhu, Z. X. Cal, R. C. Budhani, M. Suenaga, and D. O. Welch, " Structure and Effects of Radiation Damage in Cuprate Superconductors Irradiated with Several-Hundred-MeV Heavy Ions", Phys. Rev. B 48, 6436 (1993).

[19]D.X. Huang, Y. Sasaki, and Y. Ikuhara, "Influence of Ion Velocity on the Damage efficiency in the Single Ion-Target Irradiation System: Au-$Bi_2Sr_2CaCu_2O_x$", Phys. Rev. B (59) 3862-3869 (1999).

[20]L. Civale et al., "Defect Independence of the Irreversibility Line in Proton-Irradiated Y-Ba-Cu-O Crystal", Phys. Rev Lett. (65) 1164-1167 (1990).
[21]J.F. Ziegler, J.P. Biersack, V. Littmork, "The Stopping and Range of Ions in Solids", Vol. 1, Pergamon Press, Oxford, 1985.
[22]P. Poole, "Handbook of Superconductivity", Academic Press, 2000
[23]B.D. Weaver, M.E. Reeves, D.B. Chrisey, G.P. Summer, W.L. Olson, M.M. Eddy, T.W. James, and E.J. Smith, "Critical Current enhancement in Proton-Irradiated $Tl_2CaBa_2Cu_2O_8$", J. Appl. Phys. 69 (2), 1119-1121 (1991).
[24]A. Gandini, R. Weinstein, Y. R. Ren, R. P. Sawh, D. Parks, Y. C. Guo, B. Zeimetz, S. X. Dou, S. Tönies, C. Klein and H. W. Weber, " Critical Current Enhancement in $(Bi,Pb)_2Sr_2Ca_2Cu_3O_{10}$ Tapes Via Isotropic Quasi-Columnar Defects, Induced by Fission Products". Physica C 341-348,1453-1454 (2000).
[25]B. M. Vlcek et al., "Enhancements of the Critical Current of YBCO Single Crystals by Neutron and Proton Irradiation", IEEE Transaction on Appl. Supercon., Vol. 3 No. 1 (1993).

Fabrication of High Temperature Superconduct

EFFECT OF Y_2BaCuO_5 MORPHOLOGY AND SIZE IN SEMISOLID MELT ON GROWTH RATE OF $YBa_2Cu_3O_{7-x}$ SINGLE CRYSTALS

Oratai Jongprateep
Department of Ceramic Engineering
University of Missouri-Rolla
Rolla, MO 65401

Fatih Dogan
Department of Ceramic Engineering
University of Missouri-Rolla
Rolla, MO 65401

ABSTRACT

To control the growth rate of melt textured $YBa_2Cu_3O_{7-x}$ (Y123) single crystals, the effects of starting powder composition and heating rate of the compacts on the morphology and size of Y_2BaCuO_5 (Y211) in semisolid melt were studied. The processing route of the samples involved the mixture of Y_2O_3 and liquid phase composition (barium cuprate and copper oxide) as starting compounds. The correlations between the heating rate, the morphology of Y211, and the growth rate of Y123 ware examined. Observations from crystal size and microstructural analysis revealed that lower heating rate of compacts led to formation of needle shaped Y211 particles, which consequently resulted in faster growth rate of melt textured Y123 single crystal.

INTRODUCTION

The process of top seeded melt texture growth reduces weak-link problems, and greatly enhances the critical current density of Y123 samples through grain alignment. While samples with large critical current density can be fabricated, obtaining large size and high quality Y123 crystals can be limited. To optimize the size of a single crystal, it is crucial to understand growth mechanism and factors that affect the crystal growth in the fabrication process of melt textured Y123. One of the crucial factors that plays an important role in the kinetics of Y123 crystal growth is the size and morphology of Y211 inclusion in the melt. It has been shown that the controlled Y211 morphology leads to the enhancement of Y123 crystal growth [1-5].

However, the effect of some other factors such as heating rate along with powder-processing route, have not been well established. In this study, the effect of heating rate along with a composite powder processing technique impacting on Y211 morphology and crystal growth of Y123, will be presented.

EXPERIMENTAL

The Y123 pellets were prepared by solid liquid melt growth process. Samples were prepared by mixing commercially available Y_2O_3, CuO and barium cuprate (prepared by combustion synthesis). The starting compositions were pure Y123, Y123+16.7 mol% excess Y211 and Y123+23.1 mol% excess Y211. Very fine platinum powder in the amount of 0.1 wt% was added to all of the powders. The mixtures of powder were uniaxially pressed to obtain 1-inch diameter pellets. $SmBa_2Cu_3O_{7-x}$ (Sm123) was used as the seed on each pellet.

Each set of pellets was heated at a different heating rate—47°C/hr, 94°C/hr and 172°C/hr—from room temperature to 1050°C. Samples were subsequently held at 1050°C for 1 hr and cooled down rapidly to 1010°C. The cooling rate from 1010°C to 990°C remained constant at 0.286°C/hr for all sets. Samples were held at 990°C for 10 hrs prior to cooling to room temperature. Y211 morphology and size were observed using quenched samples in semisolid melt.

Crystal size of samples was determined by measuring a distance from corner to corner of the crystal growth region (Fig. 1a). For samples with the crystal region extended to the edge of the pellet, the diameter of the pellet was used as the crystal size (Fig. 1b). The microstructure of the samples were studied by JEOL-T330A scanning electron microscope (SEM).

RESULTS

Effect of Heating Rate on Crystal Growth:

For the sample with the smallest growth region, the crystal size was 0.4 cm by 0.4 cm. The crystal growth regions accounted for more than 80% of the area of the pellet surface in most of the pellets. Some samples demonstrated the growth region extended to the edge of the pellet.

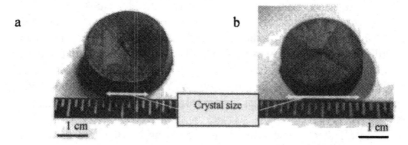

Figure 1 (a) and (b). Top surface views for samples prepared at different heating rate 172°C/hr and 47°C/hr respectively

Fabrication of High Temperature Superconduct«

Fig. 1 (a) and (b), reveals the effect of heating rates on the crystal size. For the samples with identical amount of Y211 inclusions, the lower heating rate tended to favor the crystal growth.

As shown in Figs. 2 and 3, the crystal sizes of samples were in the range of 4 mm–21 mm. The smaller crystal sizes were observed in samples with low amounts of Y211, heated at high heating rate. The larger crystal sizes were observed in samples with high Y211 content, heated at the lowest heating rate.

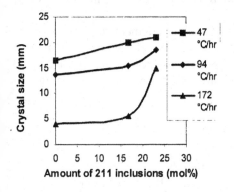

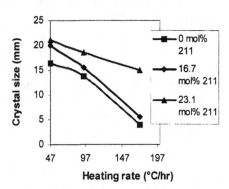

Figure 2. Growth dimension (crystal size) of Y123 as a function of Y211 inclusion

Figure 3. Growth dimension (crystal size) of Y123 as a function of heating rate

Microstructural Development:

From Fig. 4, it was observed that different heating rates resulted in different morphology of Y211 particles in semisolid melt. The shape of Y211 particles in samples with higher heating rates appeared to be platelet-like (4a), while needle-like particles were observed in the samples with lower heating rate (4b).

Figure 4 (a) and (b). SEM micrographs of Y123 samples with 23.1 mol% of Y211 inclusion, heated with heating rate 172°C/hr and 47°C/hr respectively

DISCUSSION

Mechanisms in the formation of platelet and needle-shaped 211 particles in this experiment are proposed as follows:

During the heating of powder compacts with a high heating rate, the starting compounds bypass the formation of Y123, which is the stable phase at temperatures below 1000°C. Instead, the formation of Y211 particles occurs in the melt. The direct formation of Y211 from Y_2O_3 phase results in the coarsened platelet-shaped Y211 particles (Fig. 4a).

For samples with low heating rate, an additional intermediate reaction occurs below the peritectic temperature. During the heating of samples, the starting compounds initially form Y123 phase. However, as the heating of the sample progresses, peritectic decomposition of the Y123 occurs, resulting in the formation o f Y 211 p articles a nd t he l iquid p hase. T he p rocess l eads t o n eedle-shaped Y211 particles (Fig. 4b).

The reactions involved in the described processes are shown as:

High heating rate: $Y_2O_3+BaCuO_2+CuO \rightarrow Y_2BaCuO_5 + L$

Low heating rate: $Y_2O_3+BaCuO_2+CuO \rightarrow 2YBa_2Cu_3O_{7-x} \rightarrow Y_2BaCuO_5 + L$

The advantage of the needle-shaped Y211 particles is that they can redissolve relatively fast in the liquid phase as compared to bulky platelet shape Y211. This may lead to a more rapid supply of yttrium ions in the liquid phase for the formation of Y123 resulting in faster crystal growth.

SUMMARY

The relationship between the heating rate, microstructure and crystal growth of Y123 was investigated using Y_2O_3, CuO and barium cuprate as starting powders. It was observed that samples with a low heating rate resulted in a needle-like morphology of Y211 particles in semisolid melt and in a higher solidification rate during melt textured growth of Y123 single crystals.

ACKNOWLEDGEMENT

This work was supported by a grant from the Materials Research Center at the University of Missouri-Rolla.

REFERENCES

[1]M. Murakami, A. Kondoh, H. Fujimoto, K. Takamuku, N. Nakamura, and N. Koshizuka, "A Comparison of Melt Processes to Prepare $YBa_2Cu_3O_{7-x}$ with High J_c," *Advances in Superconductivity*, 4 451–54 (1991).

[2]C.Varanasi and P.J. McGinn, "Y_2BaCuO_5 Particle Coarsening During Melt Processing of $YBa_2Cu_3O_{7-x}$," *Journal of Electronic Materials*, **22** [10] 1251–57 (1993).

[3]I. Monot, J. Wang, M.P. Delamare, J. Provost, and G. Desgardin, "Influence of Different Melt Process upon the Microstructure and Critical Current of Textured Y123," *Physica C*, **267** 173–82 (1996).

[4]Y. Shiohara, and A. Endo, "Crystal Growth of Bulk High-T_c Superconducting Oxide Materials," *Materials Science and Engineering*, **R19** [1-2] 1-86 (1997).

[5]M. L. Griffith, J. W. Halloran, and R. T. Huffman, "Formation and Coarsening behavior of Y_2BaCuO_5 from Peritectic Decomposition of $YBa_2Cu_3O_{7-x}$," *Journal of Materials Research*, **9** [7] 1633 (1994).

Fabrication of High Temperature Superconduct

FLUX PINNING and PROPERTIES OF SOLID-SOLUTION (Y,Nd)$_{1+x}$Ba$_{2-x}$Cu$_3$O$_{7-\delta}$ SUPERCONDUCTORS PROCESSED IN AIR and PARTIAL OXYGEN ATMOSPHERES

T. J. Haugan, J. M. Evans, J. C. Tolliver, I. Maartense, P. N. Barnes
Air Force Research Laboratory, 2645 Fifth St. Ste. 13, Wright-Patterson AFB, OH 45433-7919

W. Wong-Ng, L. P. Cook, R.D. Shull
National Institute of Standards and Technology, Materials Science and Engineering Laboratory, Gaithersburg, MD 20899-8520

ABSTRACT

The effect of chemical composition substitutions on the flux pinning and physical properties of (Y,Nd)$_{1+x}$Ba$_{2-x}$Cu$_3$O$_{7-\delta}$ superconductors was studied in powders processed by solid-state reaction and equilibrated in air at 910°C. The powders were subsequently processed in 1% O$_2$ atmosphere at < 920°C to increase the superconducting transition temperature (T$_c$) and critical current density (J$_c$). After processing in air, the powders were nearly pure single-phase compositions as determined by X-ray diffraction. Powders were finally annealed in 100% O$_2$ atmosphere at temperatures < 500 °C to maximize T$_c$. The T$_c$s of the powders were measured by ac susceptibility and dc magnetization methods. Annealing powders with a final step in 1% O$_2$ atmosphere compared to processing in air significantly enhanced T$_c$ from 65-90 K to > 92 K for all compositions tested, and also increased J$_c$ from about ~10^3-10^5 A/cm^2 to ~10^6 A/cm^2. The flux pinning properties varied depending on exact composition, and the intrinsic behaviors changed with the final 1% O$_2$ annealing treatment.

INTRODUCTION

High transition temperature (Rare-Earth)Ba$_2$Cu$_3$O$_{7-z}$ (RE123) superconductors are being considered for applications including thin film coated conductors and bulk devices because of their high superconducting transition temperatures (T$_c$) > 90 K, and enhanced critical current density (J$_c$) at 77 K in useful magnetic fields. While these materials have desirable qualities at 77 K, it is of interest to increase the $J_c(H)$ properties even further by increasing the flux pinning properties of the superconductor. Different methods can be considered to introduce flux pinning defects into the superconductors, including irradiation and addition of second-phase defects or precipitates. One method being considered for bulk applications is to substitute rare-earth cations into the RE123 compound. Different variations of this theme have been studied, including substitution in (Y,R)123 with R = Ho, Dy, Gd, Eu, and Pr, and various other combinations of rare-earths such as (Gd,Sm,Eu)123 [1-16]. Possible mechanisms by which such substitutions increase the flux pinning include: (a) addition of second-phase defects by precipitation or composition changes, (b) formation of finely distributed lower T_c components from the mixed solubility of RE with Ba and intersolubility of RE or other mechanisms, or (c) randomly distributed oxygen-deficient zones which have lower T_cs [7,8]. The finely distributed lower T_c components are suggested as a cause of the so-called 'fishtail' effect, where as the magnetic field is increased, the J_c increases as the lower T_c components transition to normal behavior before the J_c decreases at much higher applied magnetic fields [7,8]. The peak of J_c maximum in these materials typically occurs at applied fields of about 2 T to 3 T.

Studies of the system $(Y_{1-x}Nd_x)Ba_2Cu_3O_{7-\delta}$ have, to our knowledge, been limited thus far to melt-processed or single crystal materials [9-14], and powders [15,16]. Single crystals in this system demonstrated very high J_c and the 'fishtail' effect for Nd content varying from 0.1 to 0.4 after varying oxygenation treatments [13]. Recently our group published initial studies on $(Y_{1-x}Nd_x)Ba_2Cu_3O_{7-\delta}$ powders processed in air [16].

In this paper, previous studies on powders in this system [16] are continued by adding an additional processing step of annealing in reduced oxygen atmospheres to enhance T_c and J_c. The T_c in particular was noted to be quite low for some of the Nd-rich and Ba-poor compositions after annealing in air [16]. Subsequent processing in reduced oxygen atmospheres is well known to sharpen the T_c transition of Nd123 type compositions and improve superconducting properties [15]. How effective this is for processing the mixed compositions and their effect on flux pinning is studied herein.

In this and previous work, the effect of composition changes in $(Y,Nd)_{1+x}Ba_{2-x}Cu_3O_{7-\delta}$ was studied in solid-state powders annealed in air to achieve chemical equilibrium [16]. Powder compositions are expected to have different properties than (Y,Nd)123 melt-processed or crystals in previous studies, where non-equilibrium processes can affect the physical properties. In melt-processed materials it is almost impossible to eliminate the formation of second-phase defects such as RE211, which can affect pinning. With powder processing, it is possible to completely eliminate second-phase defects, and investigate other causes of pinning more related to the intrinsic properties of the crystal structure.

For the studies herein, powders were annealed in 1% oxygen atmosphere which is expected to enhance site occupation of Nd on the RE rather than Ba site [15,17]. This is expected to sharpen and increase the T_c transition, however i

Fabrication of High Temperature Superconducto

could decrease flux pinning by reducing the formation of lower T_c solid-solutions [8,15,16].

In this paper, only solid-solution single-phase compositions are considered that preserve the single-crystal structure however perturb the superconductor in localized regions surrounding the chemically substituted volumes. It's expected that with Nd substitution on about 5 % or 10% of the Y or Ba sites, the defect density of the locally-perturbed areas can be high enough for effective pinning ~ 2-3 $\times 10^{11}$ cm^{-2} equivalent to a fluxon density of about 3-6 T. How effectively this can increase the flux pinning must be determined, however.

EXPERIMENTAL**

Experimental methods for this work are described in the following, using procedures identical to previous work [16]. Superconducting powders were prepared by the solid-state reaction method, using starting reactants of Nd_2O_3, Y_2O_3, $BaCO_3$, and CuO ($\geq$ 99.95% purity). The powders were dehydrated at 450 °C prior to weighing. The powders were mixed and ground with mortar and pestle, calcined by slow heating 650 °C to 850 °C at 25 °C/h, and subsequently annealed with intermediate grinding at 880 °C and 910 °C. Powders were annealed at 910 °C with intermediate grinding until phase equilibrium was reached (3-4 annealings), as determined by X-ray diffraction (XRD). The powders were reacted in ~1 cm diameter pellets (0.5 g to 1 g batches), formed by lightly pressing (5-10x10^6 Pa) in molds. X-ray diffraction was performed with a Rigaku diffractometer. A step size of 0.03° was used for the θ-2θ scans.

After single-phase compositions were obtained, an additional processing step in 1% O_2 atmosphere was performed; the annealing temperature was at 820°C to 920°C increasing in 20°C increments as the Nd content increased from 0 to 1.0, and pellets were reacted on mixed (Nd,Y)211 pellets to minimize reactions with the substrate. After annealing in 1% O_2 atmosphere, the samples were checked with XRD to determine if the single phase composition was preserved or whether some slight (melting or decomposition) reaction might have occurred to cause formation of Y_2BaCuO_5 and $BaCuO_2$ phases (for example). A final annealing step was performed in 100% O_2 atmosphere at 460°C to 260°C, decreasing in 40°C increments as the Nd content increased from 0 to 1.0; this assumed a linear gradient in temperature between optimized annealing temperatures for Y123 and Nd123 compositions [18]. The optimal annealing temperature in 100% O_2 atmosphere for the mixed RE123 compounds to maximize oxygen content and T_c is unknown yet, therefore this approximation was used.

Superconducting properties of powders were measured with a SQUID magnetometer (Quantum Design, MPMS/MPMS^2). Magnetization-applied field (M-H) hysteresis loops at different temperatures were made by heating samples to 100 K and zero-field cooling (ZFC) to the measurement temperature. The magnetic J_c was estimated using the extended Bean critical current model $J_c = 15(\Delta M)/R$ where ΔM is the volume magnetization, and R is the radius of the superconducting volume roughly approximated as 0.00005 cm for the finely reacted and ground powders [19].

Field-cooled (FC) Meissner and zero-field-cooled (ZFC) measurements were performed from 5 K to 100 K [19]. The SQUID magnet was reset to zero before any measurements. The superconducting volume percentages were calculated using $\chi(\%) = 4\pi\chi_v/(1-D*4\pi\chi_v)$, where $\chi_v = M/H_{appl}$ is the measured magnetic susceptibility, and $D = 1/3$ is the demagnetization factor assuming a spherical particle distribution [19]. The applied magnetic field was $H_{appl} = 796$ A/m - H_{rem}, where H_{rem} is the remnant field of the magnet after resetting to zero, determined for each sample by measuring M(H) from 796 A/m to -398 A/m at 79.6 A/m intervals and plotting when $M = 0$ ($\pm$ 8 A/m accuracy). The transition temperature of the largest volume fraction of powder was determined by finding the temperature when $(d^2\chi/dT^2) \cong 0$ upon cooling from 100 K. The standard uncertainty of T_c measurements was 0.5 K, as determined from both ZFC and FC curves and multiple measurements. The standard uncertainty of Meissner volume fractions was $\pm$ 5%, as determined from multiple measurements of one sample. The standard uncertainty of J_cs was measured using differences of ΔM for both positive and negative applied fields. Standard uncertainties for normalized $J_c(H)$ curves were calculated by measuring errors of ΔM for positive and negative applied fields, and averaging the ΔM errors over the range of approximately 0.1 T to 4.8 T.

The superconducting transition temperature (T_c) was also measured using an AC susceptibility technique with the amplitude of the magnetic sensing field, h, varied from 0.025 Oe to 2.2 Oe, at a frequency of approximately 4 kHz. The AC susceptibility technique provides information about primary and secondary transitions of the bulk samples. T_cs were measured on palletized samples with very small thickness. Samples were mounted onto the end of a sapphire rod and measured as the samples were warmed through the transition region at very slow rate of ~ 0.06 K/min. The T_c measurements were accurate within $\leq$ 0.1 K at three calibration points: liquid He at 4.2 K, liquid N_2 at 77.2 K, and room temperature. Different methods of measuring powders were tested including mounting pellets and powders placed in silicone grease, and the different methods gave the same T_c values within about 0.1 K.

RESULTS

The range of single-phase compositions determined in previous work is shown in Figure 1 [16]. All of the compositions were nominal single-phase as determined by XRD, in agreement with previous results [15], except for a small region (Ba = 2.0 and Y = 0.3 to 0.4) where sluggish reactions were suspected [16].

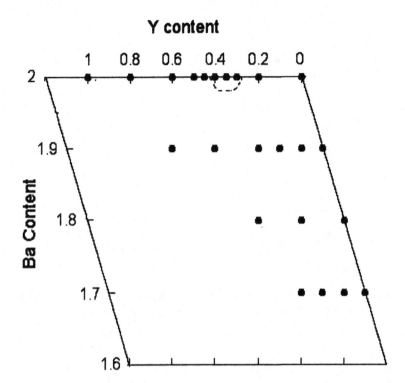

Figure 1. Compositions of $(Y,Nd)_{1+x}Ba_{2-x}Cu_3O_{7-\delta}$ that were nominal single-phase, excluding Ba = 2 and Y = 0.3 to 0.4 [16].

The superconducting transitions of Ba = 2.0 and 1.9 compositions annealed only in air determined from previous work are shown in Figure 2 [16]. The transitions were broadened for Nd > 0.4, and the transition for the largest volume fraction of the powder was reduced to ~ 73 K for the Nd = 1.0 composition. A reduction and broadening of T_c for the Nd = 1.0 composition is usually observed without further processing in reduced oxygen partial pressures [15,17]. With increasing Nd and reduced Ba content, the T_c of the bulk powder is decreased, in agreement with trends reported previously for $(Y,Nd)_{1+x}Ba_{2-x}Cu_3O_{7-\delta}$ and single-phase $Nd_{1+x}Ba_{2-x}Cu_3O_{7-\delta}$ [15,17].

After annealing these compositions in 1% reduced oxygen atmosphere, the T_cs sharpened considerably and the transition onsets were greatly increased to > 92K on average, as shown in Figure 3. The increase and sharpening of T_c for the Nd-rich compositions in reduced O_2 atmosphere is consistent with previous work [7,8,15,17]. Of special note in Figure 3 is that the T_cs of the Ba = 1.9 and Nd-rich compositions were greatly increased sufficiently high so that they can be considered for practical applications.

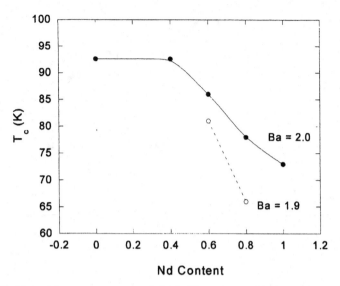

Figure 2. Transition temperature of the bulk superconducting volume fraction of powders processed in air measured by dc magnetization [16].

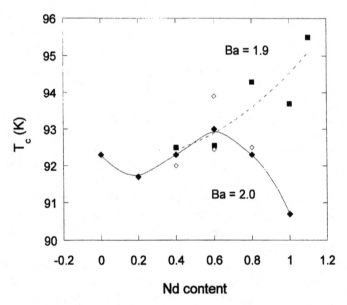

Figure 3. Transition temperature of powders processed with an additional annealing step in 1% O_2 atmosphere; Ba = 1.9 powders measured by ac susceptibility χ' onset (■) and for Ba = 2.0 powders measured by DC magnetization (◊) or ac susceptibility χ' onset (◆).

Fabrication of High Temperature Superconduct

Figure 4 plots the maximum J_cs of the powders for different composition and temperature annealed only in air. In general, the J_cs decreased with increasing Nd substitution and decrease of Ba, consistent with the decrease of T_c shown in Figure 2. Only the Y123 compound had reasonably high J_c to be considered for practical applications. The normalized $J_c/J_{c\text{-max}}$ curves for the same powders is shown in Figure 5, where $J_{c\text{-max}}$ is the maximum J_c measured in the range of H_{appl} ~ (0.06 to 0.1) T. More intrinsic differences in the pinning can be observed in Figure 5.

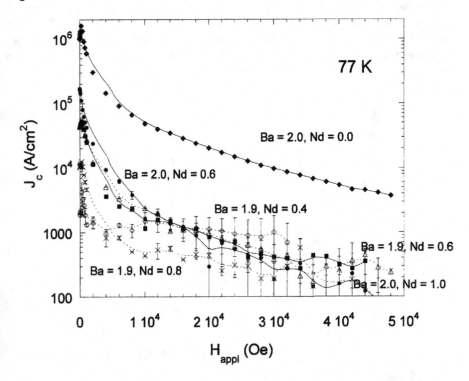

Figure 4. Critical current density as a function of applied magnetic field (H_{appl}) of powders annealed in air estimated from M-H loops for varying Nd content and Ba = 1.9 (dashed lines, open symbols) and Ba = 2.0 (solid lines, filled symbols).

The effect of the additional annealing step in 1% O_2 atmosphere is shown in Figure 6, where $J_c(H)$ curves are compared for powders processed with and without the additional step. The increase of J_c with the additional processing in 1% O_2 atmosphere can be clearly seen in Figure 6, with the J_c increased more than an order of magnitude for any applied field and all compositions. The increase is sufficiently high so that the J_cs of the Ba 1.9 compositions are about equal or better than values for Y123 material (Ba 2.0, Nd = 0), as shown in Figure 4. The

J_c of the Ba = 2.0, Nd = 0.6 compound was higher by about 50% than Y123 especially for H_{appl} > 3 T. Therefore these compositions might be considered for additional processing steps or for processing in thin film form to determine additional effects on flux pinning.

In Figure 7 the normalized J_c curves from Figure 6 are plotted, which gives indication of more intrinsic changes of pinning. For one composition (Ba = 1.9, Nd = 0.6), additional annealing in 1% O_2 atmosphere significantly changed the flux pinning behavior by enhancing the lower field J_cs (H< 2T) and reducing the higher field J_cs (H > 2T). The intrinsic pinning is better than Y123 as shown in Figure 5, which suggests that with additional optimization of $J_c(0T)$ values, the absolute values of J_c could be increased substantially also. For another composition (Ba = 1.9, Nd = 0.4), the J_c was quite low for processing in air, so the normalized curve showed more usual behavior after processing in 1% O_2 atmosphere.

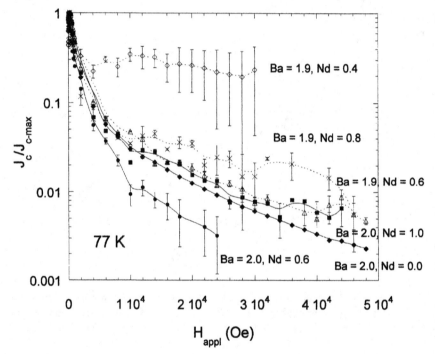

Figure 5. Normalized J_c as a function of H_{appl} for films from Figure 4; maximum J_c ($J_{c\text{-max}}$) was measured for $H_{appl} \sim$ (0.06-0.1) T.

Fabrication of High Temperature Superconduct

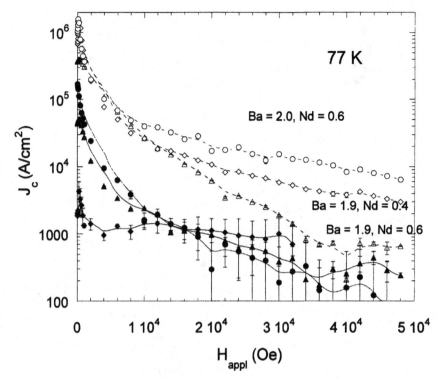

Figure 6. $J_c(H_{appl})$ comparison of powders annealed with additional step in 1% O_2 atmosphere (dashed lines, open symbols), to the same compositions annealed only in air (solid lines, closed symbols).

For all compositions 77 K, the powders showed a very small fish-tail effect, with $J_{c\text{-max}}$ occurring at 0.06 T to 0.1 T depending on composition. By comparison, the peak of maximum J_c in oxygenated melt-processed materials usually occurs around 2 to 3 T [7,8]. This suggests the powder compositions in this work are homogeneous in nature without the oxygen-deficient regions thought to cause the fish-tail effect [7,8]. A homogenous distribution of oxygen is expected as a consequence of the small size of the powders, and the near-equilibrium conditions used for processing.

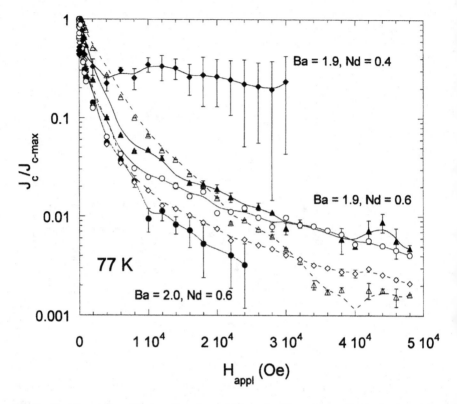

Figure 7. Normalized J_c as a function of H_{appl} for films from Figure 6; maximum J_c ($J_{c\text{-}max}$) was measured for $H_{appl} \sim (0.06\text{-}0.1)$ T. Powders were treated in 1% O_2 atmosphere (open symbols, dashed lines), or only treated in air (filled symbols, solid lines).

CONCLUSIONS

The flux pinning and superconducting properties of $(Y,Nd)_{1+x}Ba_{2-x}Cu_3O_{7-\delta}$ powders were studied for varying compositions, and compared for different final high temperature processing steps in air or 1% O_2 atmosphere. Processing of powders in air was found to form single-phase structures, however the superconductivity and T_c were much less than optimal as evidenced by low and wide T_c transitions, and J_cs were in the range of 10^3-10^5 A/cm^2 for most compositions. Subsequent processing of powders in 1% O_2 atmosphere at reduced temperatures < 920°C increased and narrowed the T_c transitions and raised J_c values to the range of 10^6 A/cm^2, where J_c values were estimated assuming a uniform distribution of spherical particles with diameter ~1.0 micron. Compositions including Ba = 2.0, 1.9 and Nd = 0.4-0.6 had sufficiently high J_c (actual values and normalized for maximum J_c) that they could be considered for

Fabrication of High Temperature Superconducto

further investigation. Annealing in 1% O_2 atmosphere was observed to change the normalized J_c pinning, e.g. by increasing J_c for H < 2 T and reducing J_c for H > 2 T for one composition (Ba = 1.9, Nd = 0.6)

For the range of single-phase compositions studied, additional processing in 1% O_2 atmosphere significantly increased T_c > 92 K for all compositions tested, including Ba = 1.9. By comparison, T_cs of the same powders processed in air were only in the range of 65-90 K, which was too low to consider for applications at 77 K. Additional processing in 1% O_2 atmosphere significantly changed the T_cs and J_cs and therefore the range of compositions than could be considered for possible application in coated conductors.

ACKNOWLEDGEMENTS
The authors would like to thank T. Spry, M.E. Fowler, and K. Fields at AFRL for assistance with sample preparation, and R. Drew of NIST-Gaithersburg for SQUID measurements.

** Certain commercial equipment, instruments, or materials are identified in this paper in order to specify the experimental procedure adequately. Such identification is not intended to imply recommendation or endorsement by the Air Force Research Laboratory or the National Institute of Standards and Technology, nor is it intended to imply that the materials or equipment identified are necessarily the best available for the purpose.

REFERENCES
[1]Y. Feng, A. K. Pradhan, Y. Zhao, Y. Wu, N. Koshizuka, and L. Zhou, "Influence of Ho substitution for Y on flux pinning in melt-processed YBCO superconductors," *Physica C* **357-360** 799-802 (2001).
[2]A. R. Devi, V. S. Bai, P. V. Patanjali, R. Pinto, N. H. Kumar, and S. K. Malik, "Enhanced critical current density due to flux pinning from lattice defects in pulsed laser ablated $Y_{1-x}Dy_xBa_2Cu_3O_{7-\delta}$ thin films," *Supercond. Sci. Technol.* **13** 935-939 (2000).
[3]H. H. Wen, Z. X. Zhao, R. L. Wang, H. C. Li, and B. Yin, "Evidence for lattice-mismatch-stress-field induced flux pinning in $(Gd_{1-x}Y_x)Ba_2Cu_3O_{7-\delta}$ thin films," *Physica C* **262** 81-88 (1996).
[4]Y. Li, G. K. Perkins, A. D. Caplin, G. Cao, Q. Ma, L. Wei and Z X Zhao, "Study of the pinning behaviour in yttrium-doped Eu-123 superconductors" *Supercond. Sci. Technol.* **13** 1029-1034 (2000).
[5]H. H. Wen, Z. X. Zhao, Y. G. Xiao, B. Yin, and J. W. Li, "Evidence for flux pinning induced by spatial fluctuation of transition temperatures in single domain $(Y_{1-x}Pr_x)Ba_2Cu_3O_{7-\delta}$ samples," *Physica C* **251** 371-378 (1995).
[6]E. S. Reddy, P. V. Patanjali, E. V. Sampathkumaran, R. Pinto, "Fabrication and superconducting properties of ternary $REBa_2Cu_3O_y$ thin films," *Physica C* **366** 123-128 (2002).

[7]M. R. Koblischka, M. Muralidhar, M. Murakami, "Flux pinning sites in melt-processed $(Nd_{0.33}Eu_{0.33}Gd_{0.33})Ba_2Cu_3O_y$ superconductors," *Physica C* **337** 31-38 (2000).

[8]M. Jirsa, M. R. Koblischka, T. Higuchi, M. Muralidhar, M. Murakami, "Comparison of different approaches to modelling the fishtail shape in RE-123 bulk superconductors," *Physica C* **338** 235-245 (2000).

[9]C. Varanasi, P. J. McGinn, H. A. Blackstead, and D. B. Pulling, "Nd Substitution in Y/Ba Sites in Melt Processed $YBa_2Cu_3O_{7-\delta}$ Through Nd_2O_3 Additions," *Journal of Electronic Materials,* **24** [12] 1949-1953 (1995).

[10]D. N. Matthews, J. W. Cochrane, G. J. Russell, "Melt-textured growth and characterization of a $(Nd/Y)Ba_2Cu_3O_{7-\delta}$ composite superconductor with very high critical current density," *Physica C* **249** 255-261 (1995).

[11]P. Schätzle, W. Bieger, U. Wiesner, P. Verges and G. Krabbes, "Melt Processing of (Nd,Y)BaCuO and (Sm,Y)BaCuO composites," *Supercond. Sci. and Technol.* **9** 869-874 (1996).

[12]A. S. Mahmoud and G. J. Russell, "Large crystals of the composite Y/Nd(123) containing various dopants grown by melt-processing in air," *Supercond. Sci. and Technol.* **11** 1036-1040 (1998).

[13]X. Yao, E. Goodilin, Y. Yamada, H. Sato and Y. Shiohara, "Crystal growth and superconductivity of $Y_{1-x}Nd_xBa_2Cu_3O_{7-\delta}$ solid solutions," *Applied Superconductivity* **6** [2-5] 175-183 (1998).

[14]D. K. Aswal, T. Mori, Y. Hayakawa, M. Kumagawa, "Growth of $Y_{1-z}Nd_zBa_2Cu_3O_x$ single crystals," *Journal of Crystal Growth* **208** 350-356 (2000).

[15]H. Wu, K. W. Dennis, M. J. Kramer, and R. W. Mccallum, "Solubility Limits of $LRE_{1+x}Ba_{2-x}Cu_3O_{7+\delta}$," *Applied Superconductivity* **6** [2-5] 87-107 (1998).

[16] T. J. Haugan, M. E. Fowler, J. C. Tolliver, P. N. Barnes, W. Wong-Ng, L. P. Cook, "Flux Pinning and Properties of Solid-Solution $(Y,Nd)_{1+x}Ba_{2-x}Cu_3O_{7-\delta}$ Superconductors", in *Processing of High Temperature Superconductors*, Ceramic Transactions Vol. 104, edited by A. Goyal, W. Wong-Ng, M. Murakami, J. Driscoll (American Ceramic Society, Westerville OH, 2003), p. 299 - 307.

[17]R. W. McCallum, M. J. Kramer, K. W. Dennis, M. Park, H. Wu, and R. Hofer, "Understanding the Phase Relations and Cation Disorder in $LRE_{1+x}Ba_{2-x}Cu_3O_{7+\delta}$," *J. of Electr. Mater.* **24** [12] 1931-1935 (1995).

[18] J. Shimoyama, Univ. of Tokyo, private communication.

[19]*Magnetic Susceptibility of Superconductors and Other Spin Systems*, Plenum Press, New York, 1991.

PHASE RELATIONS IN THE BaO-R_2O_3-CuO_x SYSTEMS

W. Wong-Ng, L.P. Cook, and J. Suh
Ceramics Division
Materials Science and Engineering Laboratory
National Institute of Standards and Technology
Gaithersburg, MD 20899

ABSTRACT

Flexible second-generation coated conductors are based on barium-yttrium-copper-oxide and barium-lanthanide-copper-oxide materials. Processing of these conductors requires phase diagram data collected under controlled-atmosphere conditions. A special experimental procedure was used for preparing BaO starting material and for the handling and heat-treatment of samples. The phase diagrams of the carbonate-free BaO-R_2O_3-CuO_x (where R=Nd, Sm, and Y) were constructed under atmospherically-controlled conditions. A discussion of the crystal chemistry and phase relationships of selected phases found in these systems is presented. A trend as a function of the size of R was observed.

INTRODUCTION

Phase diagrams are 'road maps' for processing of $Ba_2RCu_3O_{6+z}$ materials (R-213, where R=lanthanides and Y). Phase diagrams are especially important for processing routes based on the melt-growth method [1, 2]. A number of studies pertaining to the BaO-R_2O_3-CuO_x phase equilibria have been reported. A majority of the data was collected using $BaCO_3$ as a starting reagent [1, 3-6]. Other reported Ba-containing starting reagents include BaO_2 [7], BaO and $Ba(NO_3)_2$ [8]; however, most of the reported data was not collected entirely under controlled-atmosphere conditions. As a result, the presence of carbonate or hydroxide in the samples has rendered phase diagram determinations in the high-BaO region tentative, and the reported diagrams are incomplete.

This paper is a part of a continuing effort to understand the phase equilibria and crystal chemistry of AO-R_2O_3-CuO_x systems (A = alkaline-earths). We will concentrate on three systems (A=Ba, and R = Nd, Sm and Y) that were prepared under controlled atmosphere using BaO-derived starting materials. Particular emphasis is placed on the region near the BaO corner, and on the solid solution formation in $Ba_{2-x}R_{1+x}Cu_3O_{6+z}$ and its tie-line relationships with neighboring compounds. This work has utilized special experimental procedures to minimize the presence of carbonate and minimize exposure to atmospheric CO_2 and moisture. Experimental work was carried out at two oxygen pressures, namely, at $p_{O2} = 100$ Pa (0.1 % O_2 in Ar volume fraction) and at $p_{O2} = 21$ kPa (21 % O_2 in Ar volume fraction). The former p_{O2} was selected to approximate ion beam assisted deposition (IBAD) and rolling assisted biaxially textured substrate (RABiTS) processing conditions.

EXPERIMENTAL[1]

The primary issue in our sample preparation procedure was controlling the atmosphere to avoid CO_2 and moisture. A reducing atmosphere during heat-treatment is also important to match the deposition conditions of the R-213 films. Since it is critical to use pure BaO as a starting material, we prepared BaO by decomposing $BaCO_3$ (99.99% purity, metals basis) at 1300 °C in a vacuum furnace. Sample weighing and handling were performed inside a glove box, and sample transference between the glove box and the atmosphere-controlled furnace was accomplished using a sealed transfer vessel.

A solid-state reaction technique was used for sample preparation. Since the commercial Nd_2O_3 and Sm_2O_3 powder used for this study (99.9% purity, metals basis) react readily with atmospheric moisture to form $Nd(OH)_3$ and $Sm(OH)_3$ respectively, a preliminary heat-treatment at 450 °C was necessary to convert hydroxide into oxide prior to sample preparation. After weighing and homogenization of individual compositions prepared from BaO, R_2O_3 (R=Nd, Sm and Y) and CuO (99.99 % purity, metals basis), pellets were pressed and placed inside individual MgO crucibles, and then transferred to a controlled atmosphere annealing furnace using a transfer vessel and an interlock system. Experiments were conducted at $p_{O2} = 100$ Pa and at $p_{O2} = 20$ kPa for R=Nd and Y, and at 100 Pa for R=Sm. Oxygen partial pressures were maintained by a mass flow-meter connected to cylinders of O_2 and Ar, and were monitored by a zirconia oxygen sensor at both the gas inlet and the outlet of the furnace. Heat-treatments were performed at 810 °C. Several intermediate grindings and pressings took place until the content of the phase assemblages did not change, as determined from the results of powder x-ray diffraction. A total of 281 samples were prepared for this study (142 for the Nd-systems, 35 for the Sm-system and 104 for the Y-systems).

For x-ray phase analysis, specimens were loaded into a sealed cell designed by Ritter [9]. The process of sample loading was performed inside an Ar-filled

---

[1] Certain trade names and company products are mentioned in the text or identified in illustration in order to adequately specify the experimental procedure and equipment used. In no case does such identification imply recommendation or endorsement by National Institute of Standards and Technology.

Fabrication of High Temperature Superconducto

glove box. X-ray powder diffraction was used to identify the phases synthesized and to confirm phase purity. Data were collected using a computer-controlled automated diffractometer equipped with a theta-compensation slit; CuK_α radiation was used at 45 kV and 40 mA. The radiation was detected by a scintillation counter and a solid state amplifier. A Siemens diffraction software package was used for diffractometer control and data acquisition. Phase identification was accomplished using reference x-ray diffraction patterns from the ICDD Powder Diffraction File (PDF) [10].

DISCUSSION

In the following, discussion about the phase diagrams of the BaO-$\frac{1}{2}R_2O_3$-CuO_x systems (R=Nd, Sm,Y) and a comparison of the diagrams prepared under 100 Pa oxygen partial pressure will be given. The notation adopted for lanthanide oxides is $\frac{1}{2} R_2O_3$, so that the composition ratio is the same as the cation ratio, as it is for the other end members of the system.

BaO-$\frac{1}{2}Nd_2O_3$-CuO_x Systems

(1) *Phase diagrams*

Figures 1a and 1b give the ternary phase diagrams of BaO-$\frac{1}{2}Nd_2O_3$-CuO_x prepared at p_{O2} = 21 kPa (930 °C), and p_{O2} = 100 Pa (810 °C), respectively. The diagram prepared at p_{O2} = 21 kPa is similar to that reported by Abbittasta et al. which was prepared under pure O_2 [8]. Both the solid solution $(Ba,Nd)_2CuO_{3+z}$ and the $Ba_4Nd_2Cu_2O_9$ phase were identified in the BaO-rich domain. However, many of the phases previously reported in the literature, when $BaCO_3$ was used, were not observed. For example, $Ba_3RCu_2O_z$, $Ba_4RCu_2O_z$, $Ba_5RCu_3O_z$ [11-12], $Ba_4RCu_3O_z$, $Ba_8R_3Cu_5O_z$ [13] and $Ba_6RCu_3O_z$ [6, 14] were not observed regardless of the oxygen pressure. Fu et al. [15] reported that Ba_2NdCuO_z and the "brown phase" $BaNd_2CuO_z$ are part of the same solid solution series, rather than distinct phases.

(2) *$Ba_{2-x}Nd_{1+x}Cu_3O_{6+z}$ Solid Solution*

At p_{O2} = 21 kPa, the solid solution range 'x' of the superconductor solid solution, $Ba_{2-x}Nd_{1+x}Cu_3O_{6+z}$, was confirmed to extend nominally from x = 0.0 to x = 0.95. X-ray diffraction patterns of $Ba_{2-x}Nd_{1+x}Cu_3O_{6+z}$, as quenched using flowing liquid nitrogen-cooled helium, are shown in Fig. 2. These patterns illustrate that single phase Nd-213 was obtained. Lattice parameter determination using the least-squares refinement program LSQ-NBS92 [16] indicated that samples with 0.7 > x ≥ 0 are tetragonal. Peak splitting in the x-ray patterns appears for samples with x > 0.7. We were able to index these patterns using the supercell reported by Abbattista et al. [8, 17] and Petrykin et al. [18], namely, orthorhombic Ammm, with $a'=a$, $b'=2a$ and $c'=2c$ [19]. A plot of the unit cell

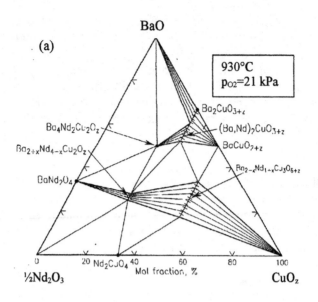

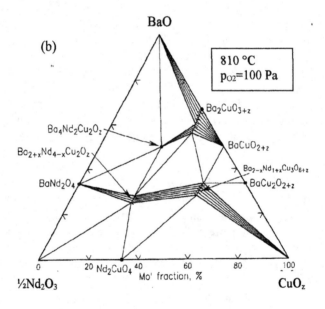

Fig. 1. Phase Diagrams for the BaO- ½Nd$_2$O$_3$-CuO$_x$ system under (a) 21 kPa and (b) 100 Pa oxygen partial pressure.

Fabrication of High Temperature Superconduct•

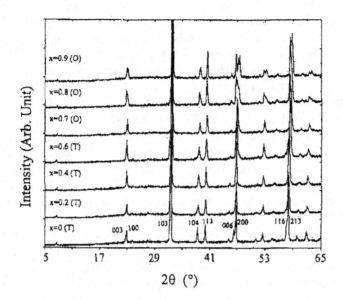

Fig. 2. X-ray diffraction patterns of the $Ba_{2-x}Nd_{1+x}Cu_3O_{6+z}$ series after Annealing at $p_{O2} = 21$ kPa at 930 °C and quenching into helium gas cooled by liquid nitrogen.

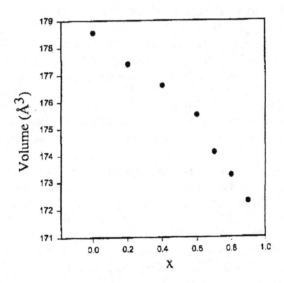

Fig. 3. Plots of unit cell volume against x in $Ba_{2-x}Nd_{1+x}Cu_3O_{6+z}$ (experiments at p_{O2}=21 kPa). For $0.95 \geq x \geq 0.7$, V of the tetragonal subcell is used.

volume, V as a function of x in $Ba_{2-x}Nd_{1+x}Cu_3O_{6+z}$ is illustrated in Fig. 3. For the orthorhombic phases, the approximate subcell volumes (one-fourth of the volume of the supercell) are used. Since the size of Nd^{3+} is smaller than that of Ba^{2+} [20], a decrease of cell volume as a function of Nd substitution is observed. Two lines with different slope are inferred from Fig. 3, one corresponding to the tetragonal phase and the other to the orthorhombic phase. For $Ba_{2-x}Nd_{1+x}Cu_3O_{6+z}$ samples that were slow-cooled at p_{O2} = 21 kPa from 900 °C, three different phases are observed in the x-ray patterns [21]. Phase transformations occurred from the orthorhombic (O), (0<x<0.2, $Pmmm$), to the tetragonal (T) phase, (0.2<x<0.7, $P4/mmm$), and then to another orthorhombic phase (O'), (0.95 ≥ x ≥ 0.7). This observation is in agreement with the results of Goodilin et al. [5]. The difference in symmetry of these slow-cooled phases relative to that of the helium-quenched samples apparently results from the different oxygen content.

The solid solution range for $Ba_{2-x}Nd_{1+x}Cu_3O_{6+z}$, prepared at p_{O2} = 100 Pa, with 0.3 > x ≥ 0, is narrower than for those prepared under at p_{O2} = 21 kPa or under oxygen. The liquid N_2/He quenched samples are tetragonal. The narrower solid solution range occurs because extra oxygen is needed to stabilize the solid solution members that are Nd-rich, and the oxygen partial pressure is not sufficient to stabilize such members [22].

BaO-½Sm₂O₃-CuOₓ system

(1) *Phase Diagram*

The diagram of the BaO-½Sm₂O₃-CuOₓ system prepared under 100 Pa 0.1% O_2 is substantially different from those prepared under air and from those prepared using $BaCO_3$ [7, 23-25]. These differences include the phase formation and the tie-line relationships. The diagram of the BaO-Sm₂O₃-CuOₓ system is somewhat similar to the Nd-system reported previously (Fig. 4), except that in the Sm-system, the "brown phase" $BaNd_2CuO_5$ (tetragonal $I4/mcm$) is replaced by the orthorhombic "green phase" $BaSm_2CuO_5$. Also, because the difference of the ionic size between Ba^{2+} and Sm^{3+} is greater than that between Ba^{2+} and Nd^{3+}, the extent of solid solution also in the Sm system is smaller. The structure of the BaR_2CuO_5 "green phase" ($Pnma$ space group) has been studied extensively [26-28].

There are a total of four phases in the ternary BaO-½Sm₂O₃-CuOₓ system Among them, two form solid solutions. These four phases are $Ba_{2-x}Sm_{1+x}Cu_3O_{6+z}$ ($Ba_{2-x}Sm_x)CuO_{3+z}$, $BaSm_2CuO_5$, and $Ba_4Sm_2Cu_2O_{9-z}$. The perovskite-related phase, $Ba_6SmCu_3O_z$ (Sm-613), although it exists when prepared in air [7], was no observed under 100 Pa O_2 (The Sm-613 composition that we prepared showed a mixture of $Ba_4Sm_2Cu_2O_{9-z}$ and (Ba, Sm)$_2CuO_{3+x}$). The cubic perovskite-related phase, $Ba_4SmCu_3O_{8.5+z}$ phase, which was reported to exist under oxygen, was also

Fabrication of High Temperature Superconduct

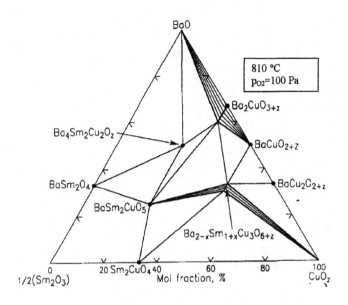

Fig. 4. Phase Diagram for the BaO-½Sm$_2$O$_3$-CuO$_x$ system under 100 Pa oxygen partial pressure.

found to be unstable under 100 Pa O$_2$. Apparently when the size of R is smaller, such as that of Y, the 413 phase is stable under 100 Pa O$_2$ [7].

(2) *Structure of Ba$_4$Sm$_2$Cu$_2$O$_{9-z}$*

Under 100 Pa p$_{O2}$, Ba$_4$Sm$_2$Cu$_2$O$_{9-z}$ was found to be orthorhombic with a space group *Pnn2* (a subgroup of *P4n2*, No. 34). The (001) projection of the Ba$_4$Sm$_2$Cu$_2$O$_{9-z}$ structure is shown in Fig. 5a and the one-dimensional chains of CuO$_5$ are shown in Fig. 5b. The lattice parameters determined using powder x-ray diffraction are $a = 11.9718(7)$X, $b = 11.8884(7)$ X, $c = 3.8465$ (2) X, and $V = 547.47$X^3). This semi-conductor phase is one of the few cuprate phases reported which possesses the unusual 1-dimensional chains of CuO$_5$ units [29]. These distorted and isolated CuO$_5$ units consist of square pyramids with four almost equal basal Cu-O distances (1.921(6) Å to 1.935(5) Å) and one much longer axial Cu-O bond of 2.473(14) Å. These square pyramids form corner-shared infinite chains running parallel to the c-axis. The CuO$_5$ chains are connected via Ba^{2+} and Sm^{3+} cations. The SmO$_7$ adopts a distorted mono - capped trigonal prismatic

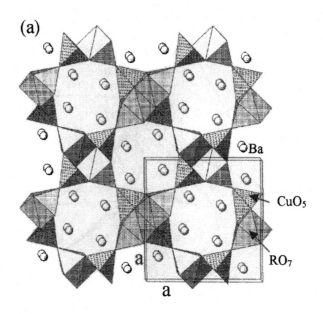

(a)

○Ba

CuO₅

RO₇

a

a

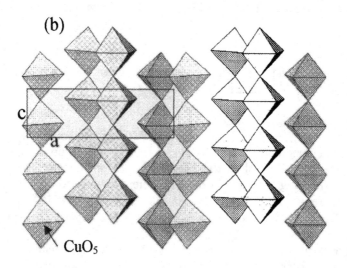

(b)

c

a

CuO₅

Fig. 5. Crystal Structure of $Ba_4R_2Cu_2O_x$ (R=Nd and Sm).
(a) The (001) projection, and (b) one-dimensional chains of CuO_5

Fabrication of High Temperature Superconduct

configuration. The environment of Sm^{3+} in Sm-422 is similar to that of the 'green phase' $BaSm_2CuO_5$ [23]. The SmO_7 and CuO_5 units form a 3-dimensional network, and wide octagonal tunnels are found throughout the structure along the c-axis inside which Ba ions are found.

$BaO-\frac{1}{2}Y_2O_3-CuO_x$ system

Figures 6a and 6b show the phase diagrams of the $BaO-Y_2O_3-CuO_x$ system prepared at $p_{O2} = 21$ kPa (900 °C) and at $p_{O2} = 100$ Pa (810 °C). The phases formed under these two conditions are similar, except for those in the binary $BaO-CuO_x$ system. For example, the reduced phase $BaCu_2O_{2+x}$ is stable when prepared under 0.1% O_2, but not under purified air. There are four ternary oxides in the $BaO-Y_2O_3-CuO_x$ system, namely, BaY_2CuO_5 (2:1:3), $Ba_4YCu_3O_x$ (4:1:3), $Ba_6YCu_3O_x$ (6:1:3) and BaY_2CuO_5 (1:2:1). The compatibility relationships of these four phases are similar in these two diagrams, but differ from those prepared using $BaCO_3$ [7, 11, 30], particularly in the Ba-rich region. The occurrence of the compounds $Ba_4YCu_3O_x$ and $Ba_6YCu_3O_x$ in the BaO-rich part of the diagram agrees with that reported by Abbattista et al. [31] and Osamura and Zhang [7]. The $Ba_2YCu_3O_{6+x}$ phase is tetragonal P4/mmm when quenched from 900 °C, and 810 °C.

A tie-line was found between $Ba_4YCu_3O_x$ and $Ba_2YCu_3O_x$, which is in contrary to the literature reported tie-line between $BaCuO_{2+x}$ and BaY_2CuO_5. In the relatively Cu-poor regions, while a tie-line was found between BaY_2CuO_5 and $Ba_6YCu_3O_x$ under pure air, $Ba_3Y_4O_9$ and $Ba_4YCu_3O_x$ were found to be compatible with each other under 0.1% O_2.

Comparison of the $BaO-\frac{1}{2}R_2O_3-CuO_x$ (R=Nd, Sm and Y) diagrams

As a comparison, Figs. 7(a) to 7(c) give the three diagrams of the Ba-R-Cu-O systems prepared under 100 Pa. A trend in phase formation is observed. Systems with larger size of R show a greater number as well as a larger range of solid solutions. For example, the R=Nd and Sm diagrams show solid solutions such as $Ba_{2-x}Nd_{1+x}Cu_3O_{6+z}$, and $(Ba_{2-x}Sm_x)CuO_{3+z}$, while in the R=Y system, the size of Y is smaller and no solid solution is observed. The 422 phase that can be prepared in the Nd and Sm systems is absent in the Y system. In the $Y_2O_3-CuO_x$ binary system, the $Y_2Cu_2O_5$ phase is formed instead of R_2CuO_4 as found in the $\frac{1}{2}Nd_2O_3-CuO_x$ and $\frac{1}{2}Sm_2O_3-CuO_x$ systems. While the literature reports a tie-line between $BaCuO_{2+x}$ and BaR_2CuO_5 [7,11,23], this tie-line is not found in the carbonate-free $BaO-R_2O_3-CuO_x$ systems.

ACKNOWLEDGEMENTS

Partial financial support from the US Department of Energy is acknowledged. The authors would like to thank Nils Swanson for his graphical assistance.

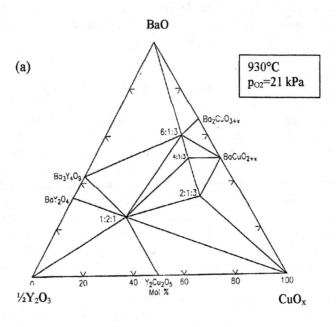

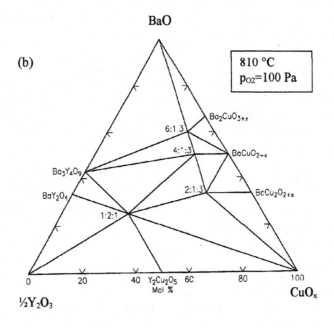

Fig 6. Phase diagrams of the BaO- ½Y$_2$O$_3$-CuO$_x$ systems Prepared under 21 KPa and 100 Pa oxygen partial pressure.

Fabrication of High Temperature Superconduct•

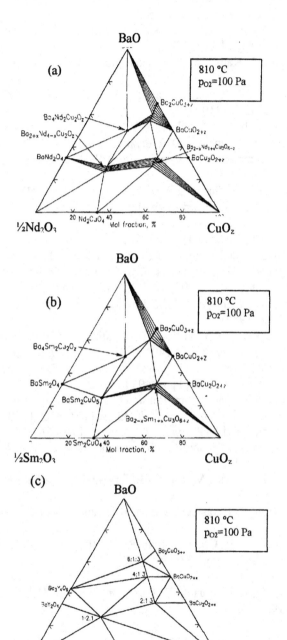

Fig. 7. Phase diagrams of the BaO- ½ R_2O_3-CuO systems, R=(a) Nd, (b) Sm and (c) Y. A trend of phase formation is observed.

REFERENCES

[1]M. Kambara, T. Umeda, M. Tagami, X. Yao, E.A. Goodilin, and Y. Shiohara, *J. Am. Ceram. Soc.* **81** (8), 2116 (1998).

[2] K. Oka, M. Saito, M. Ito, K. Nakane, K. Murata, Y. Nishihara, and H. Unoki, *Jpn. J. Phy. Lett.* **28** (2) L219 (1989)

[3]E. Goodilin, M. Kambara, T. Umeda and Y. Shiohara, *Physica C* **289**, 251 (1997).

[4]W. Wong-Ng, L.P. Cook, B. Paretzkin, M. D. Hill and J.K. Stalick, *J. Am. Ceram. Soc.* 77(9), 2354 (1994).

[5]E.A. Goodilin, N.N. Oleynikov, E.V. Antipov, R.V. Shpanchenko, G. Yu. Popov, V.G. Balakirev, and Yu. D. Tretyakov, *Physica C* **272**, 65 (1996).

[6]H. Wu, M.J. Kramer, K.W. Dennis, and R.W. McCallum, *Physica C* **290**, 252 (1997).

[7]K. Osamura and W. Zhang, *Z. Metallkd.* **84** (8) 522 (1993).

[8]F. Abbattista, D. Mazza and M. Vallino, *Eur. J. Solid State Inorg. Chem.* **28**, 649 (1991).

[9] J.J. Ritter, *Powd. Diff.* **3** (1), 30 (1988).

[10]PDF, Powder Diffraction File, produced by International Centre for Diffraction Data (ICDD), 12 Campus Blvd., Newtown Square, PA. 19073-3273.

[11]R.S. Roth, C.J. Rawn, F. Beech, J.D. Whitler, and J.O. Anderson; pp. 13-26 in *Ceramic Superocnductors II*. Edited by M.F. Yan, American Ceramic Society, Wersterville, OH, (1988).

[12]S.N. Koshcheeva, V.A. Fotiev, A.A. Fotiev, and V.G. Zubkov, *Izv. Akad. Nauk SSSR, Neorg. Mater.* **26**[7], 1491 (1990), *Inorg. Mater.* **26** [7] 1267 (1990).

[13]D.M.Deleeuw, C.A.H.A. Mutsaers, C. Langereis. H.C.A. Smoorenburg, and P.J. Rommers, *Physica C*, **152** [1] 39-49 (1988).

[14]S.I.Yoo and R.W. McCallum, *Physica C* **210** 147-156 (1993).

[15]S.J. Fu and S.S. Xie, *Chin. Sci. Bull.* **35** [10] 816-820 (1990).

[16]D.E. Appleman and H.T. Evans, Jr., Report PB216188, U.S. Department oi Commerce, National Technical Information Service,5285 Port Royal Rd. Springfield, VA 22151.

[17]F. Abbattista, C. Brisi, M. Lucco-Borlera, and M. Vallino, *Nuovo Cimento* 1 611 (1988).

[18]V.V. Petrykin, P. Berastegui and M. Kakihana, *Chem. Mater.* **11**, 3445-345 (1999).

[19]W. Wong-Ng, L.P. Cook, J. Suh, R. Coutts, J.K. Stalick, I. Levin, and Q Huang, *J. Solid State Chem.* **173** 476 (2003).

[20]R.D. Shannon, *Acta Crystallogr.* **A32**, 751 (1976).

[21]W. Wong-Ng, C.K. Chiang, B. Paretzkin, and E.R. Fuller, Jr. *Powd. Diffr.* (1), 26-32 (1990).

[22]Y. Shiohara and E.A. Goodilin, Handbook on the Physics and Chemistry of Rare Earths, **30**, edited by K.A. Gschneidner, Jr., L. Eyring and M.B. Maple, Elsevier Science B.V., Ch. 189, pp. 67-227 (2000).

[23]W. Wong-Ng, B. Paretzkin, and E. R. Fuller, Jr., *J. Solid State Chem.* **85**, 117 (1990).

[24]Z.Y. Qiao, X.R. Xing, W.X. Yuan, S.K. Wei, X.L. Chen, J.K. Liang, and S.S. Xie, *J. Alloys Compd.*, **202** [1-2] 77-80 (1993).

[25]J. Czerwonka and H.A. Eick, *J. Solid State Chem.*, **90** [1] 69-78 (1991).

[26]R.M. Hazen, L.W. Finger, R.J. Angel, C.T. Prewitt, N. L. Ross, H. K. Mao, & C.G. Hadidiaco, P.H. Hor, R.L. Meng, C.W. Chu, *Phys. Rev.* B. **35** 7238 (1987).

[27]S.F. Watkins, F.R. Fronczek, K.S. Wheelock, R.G. Goodrich, W.O. Hamilton, and W.W. Johnson, *Acta Cryst.* **C44** 3 (1988).

[28]W. Wong-Ng, M.A Kuchinski, H.F. McMurdie and B. Paretzkin, *Powd. Diff.* **4** 1 (1989).

[29]B. Domenges, F. Abbattista, C. Michel, M. Vallino, L. Barbey, N. Nguyen, and B. Raveau, *J. Solid State Chem.* **106** 271 (1993).

[30]W. Reichelt, H. Wilhelm, G. Foersterling, and H. Oppermann, *Crysta. Res. Technol.* **24** [2] K26 (1989).

[31]F. Abbattista, M. Vallino, and D. Mazza, *Mater. Chem. Phys.* **21** 521 (1989).

Fabrication of High Temperature Superconduct

STUDIES ON NANOPARTICULATE INCLUSIONS IN Y-123 THIN FILMS

S. Sathiraju[1], P. T. Murray[2], T. J. Haugan, R. M. Nekkanti[3], L. Brunke[3],
I. Maartense[4], A.L. Campbell[4], J. P. Murphy[5], J. C. Tolliver, P. N. Barnes
[1]National Research Council, [2] University of Dayton,
[3] Universal Energy Systems Inc. [4] Materials Laboratory,
Propulsion Research and Power Generation, Air Force Research Laboratory,
Wright-Patterson Air Force Base, OH-45433

ABSTRACT

The effect of nanoparticulate inclusions in pulsed laser deposited $YBa_2Cu_3O_{7-x}$ (Y-123) superconducting thin films has been studied. Nanoparticles of Ag, Y_2BaCuO_5 (Y-211), and $YBa_2Cu_3O_x$ (YBCO) were produced in the gas phase using 50mJ of pulsed laser energy at a pressure of 1 Torr O_2 for these studies. A multilayer structure of alternating YBCO thin films and nano particles (0-8 layers) were sequentially deposited using pulsed laser deposition (PLD) method. The search for effective pinning centers using this process resulted in microstructural degradation and lower critical current densities (J_c) than the pure YBCO thin films. Our studies indicate that these nano-size pinning centers introduced using this deposition method are ineffective and resulted in a poor microstructure. Granular superconductor models of De Gennes, Clark, Ambegaokar and Bartoff were used to explain the grain boundary nature. X-ray diffraction (XRD), ac susceptibility, dc electrical transport measurements in self field, and atomic force microscope (AFM) were used to characterize the samples.

INTRODUCTION

Y-123 is a type II superconductor and does not exhibit the full Meissner effect. Because of this, quantized magnetic flux or fluxons enters into the superconductor. If these fluxons are not effectively pinned, degradation in the amount of supercurrent will occur. Finding a means to more effectively pin the magnetic flux will allow higher current densities in the material. This is particularly important to the development of the HTS coated conductor. Improvement in the HTS layer, especially Y-123, will allow higher engineering current densities for the overall structure. This will allow further reduction in the weight of HTS power components.

Because of these advantages our group is actively investigating a variety of methods for generating different flux pinning mechanisms in order to enhance the J_c in Y-123 thin films[2-4]. One means of doing of so is, to create the nanoparticles insitu by using PLD[2]. Ag, Y-211 and YBCO nanoparticles were selected for inclusion in YBCO matrix to study their pinning nature. In the following work we report the addition of these nanoparticles to Y-123 thin films deposited on single crystal $LaAlO_3$(100) substrates.

EXPERIMENTAL

The nanoparticles were produced within the vapor phase of pulsed laser deposition using laser energies of 50 mJ per pulse and 1 Torr O_2 pressure where the details of deposition are provided elsewhere[2,3]. The Initial layer of Y-123 thin film was grown for 5 minutes under conditions typically used for Y-123[3]. A 5 minute deposition of Y-123 resulted in an approximately 500 Å thick layer. Nanoparticles of Y-211, YBCO and Ag were deposited on the initial Y-123 thin film for 1 minute. Alternating depositions of the nanoparticles and Y-123 thin film were deposited sequentially with the final top layer being Y-123. The final films consist of 1 to 8 nanoparticle layers deposited between the pureY-123 layers. To determine their superconducting properties, plain YBCO thin films were deposited having the same thickness as the composite structure of the YBCO with the nanoparticulate inclusion.

For this work composite samples of consisted of 2, 4, and 8 layers of nanoparticulate inclusions as well as comparison of Y-123 films with no inclusions. Films were characterized using X-ray diffraction and DC electrical resistivity measurements in self field. Thickness of the samples was measured using a Varian thickness profilometer. Surface morphology was obtained using atomic force microscope (AFM).

RESULTS AND DISCUSSION

Highly <001> oriented films were observed for 0 and 2 layers of nanoparticulate inclusions. However, XRD peaks began to broaden considerably for 4 and 8 layers of nanoparticulate inclusions. Self field (zero applied field) critical current densities decreased from 10^6 to 10^5 A/cm^2 at 77 K with the increase in number of nanoparticulate layers when compared with that of the plain Y-123 film without nanoparticulates. Fig.1a shows the T_c vs. number of layers o nano inclusions and Fig.1b shows the J_c vs. number of layers of nano inclusions.

Fabrication of High Temperature Superconduct

The reason for the decrease in the critical current density could be due to the non stoichiometric/oxygen deficient nanoparticles of YBCO or Y-211 reacting

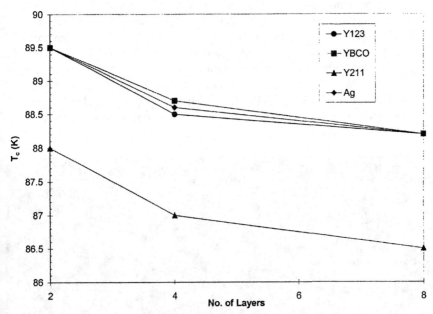

Figure 1(a). Number of layers of nanoinclusions versus critical temperature T_c

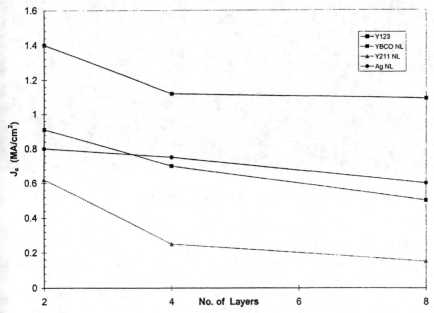

Figure 1(b). Number of layers of nanoinclusions versus J_c of composite Y-123 thin films

with the stochiometric Y-123 layers during deposition[3]. However, it is more likely caused by poor surface morphology which leads to subsequent non-epitaxial growth. Non-stoichiometry is introduced due to the low pulse energy (50mJ/per pulse) used to produce the vapor phase nanoparticles. As the number of nanoparticle layers increase, the poor growth is evident as the microstructure appears to have more voids and defects. Surface morphology of these films from AFM pictures show that the grain boundaries are widened when compared with the pure Y-123 films. Figure 2a shows the plain Y-123 film having same thickness that of the composite Y-123 film with YBCO nanoparticles. Figure 2b shows the typical AFM picture of the Y-123 composite film with 8 nano particulate layers of YBCO.

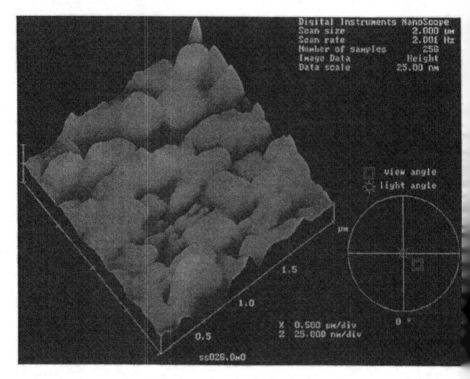

Figure 2(a). Surface morphology of pure Y123 thin film

To support the microstructural results theoretical models proposed by De Gennes and Clarke[6,7] can be invoked to get an idea about the variations in J_c as a result of changes in the microstructure. These models have successfully used earlier to explain the change in J_c due to improved microstructure in Ag doped Y-123 thin films[8]. The current flow between two superconducting grains separated by a grain boundary (weak link) caused due to the tunneling of cooper pairs is given by

Fabrication of High Temperature Superconduct

$$J_c \propto (T_c - T)^2 \exp(-d/\xi_N)$$

Where d is the thickness of the grain boundary layers and ξ_N is normal metal coherence length. Ignoring the weak temperature dependence of ξ_N compared to the $(T_c\text{-}T)^2$ term, exponential factor can be neglected at temperatures very close to T_c. Hence, very close to T_c

$$\sqrt{J_c} \propto (T_c - T)^2$$

A straight line plot of $\sqrt{J_c}$ vs. $(T_c\text{-}T)$ indicates a superconductor-normal metal-superconductor (SNS) type of network. The slope gives an idea about the extent of grain boundary width present. The higher the straight line slope indicates the smaller the grain boundary width or very well oriented grains with good grain connectivity. Figure 3a indicates the $\sqrt{J_c}$ vs. $(T_c\text{-} T)$ plots for nanoparticulate (Ag, YBCO, and Y-211) included Y-123 thin films for n= 2. If barrier between the superconducting grains are of an insulating nature, near T_c, using Ambegaokar and Baratoff[9] (AB) model one can find,

$$J_c \propto (T_c - T)$$

The cross over from one model to the other model can be observed when $\in$ = 1 a cross over takes place. Figure 3b shows the J_c vs.$(T_c\text{-}T)$ plots of Y-123 composite films for n = 4 and 8 of YBCO and Y-211 nano inclusions indicate superconductor- insulator-superconductor (SIS) kind of grain boundary nature near T_c.

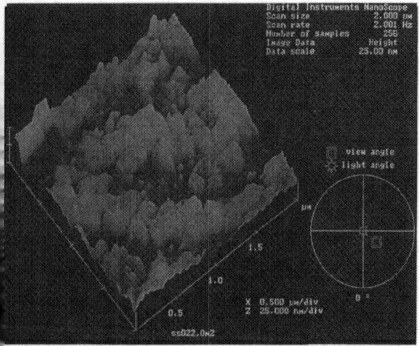

Figure 2(b). Surface morphology of the Y-123 composite film with 8 nano particulate layers of YBCO.

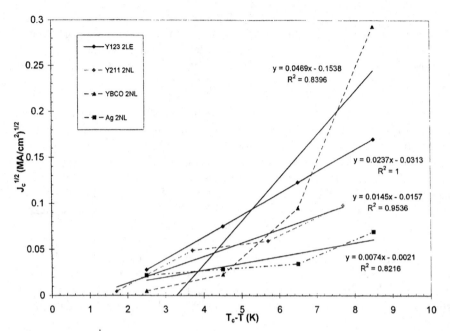

Figure 3(a). $\sqrt{J_c}$ vs. $(T_c - T)$ plots indicating SNS nature for $n = 2$ layered composite Y-123 thin films.

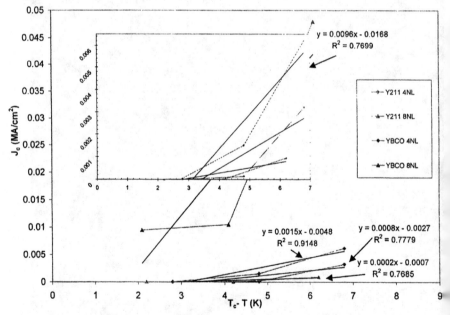

Figure 3(b). J_c vs. $T_c - T$ plots indicating SIS nature for $n = 4$ and 8 layered composite Y123 film.

Fabrication of High Temperature Superconduct

Note that the above models are valid near T_c only based on the previously stated assumptions

CONCLUSIONS

We have systematically studied the nanoparticle inclusions in Y-123 thin films using YBCO, Y-211 and Ag nanoparticles created in the vapor phase during pulsed laser deposition. The superconducting properties of the composite films have decreased with the increasing number of included nanoparticle layers. Degradation of superconducting films resulted in J_c measurements at 77 K, in self field, to drop from 10^6 A/cm^2 to 10^5A/cm^2. De –Gennes and Clarke model and Ambegaokar and Baratoff models were used to corroborate our results for the film degradation. The experimental results are in good agreement with these models. We speculate that the low laser pulse energy which was used to produce the nanoparticles resulted in poor surface morphology of the Y-123 layer. As such the composite Y-123 thin films decreased in quality.

REFERENCES

[1] M. Murakami, S. Gotoh, H. Fujimoto, K. Yumaguchi, N. Koshizuka, S. Tanaka, Supercond. Sci. Technol. 4, 543-46 (1991)

[2] P.N. Barnes, P.T. Murray, T. Haugan, R. Rogow, G.P. Perram, "Insitu creation of nano particles from YBCO " Physica C 377, 578-83 (1991)

[3] T.J. Haugan, M.E. Fowler, J.C. Tolliver, P.N. Barnes, W. Wong-Ng, and L.P. Cook, "Flux Pinning and Properties of (Y,Nd)$_{1+x}$Ba$_{2-x}$Cu$_3$O$_{7-d}$ Superconductors, Proceedings of Acers 104 Annual meeting.

[4] T.J. Haugan,, P.N. Barnes, I. Maartense, E.J. Lee, M. Sumption, and C.B. Cobb, "Island growth of Y$_2$BaCuO$_5$ nanoparticles in (211$_{\sim1.5nm}$/123$_{\sim10nm}$)xN composite multilayer structures to enhance flux pinning of YBCO films", submitted to JMR.

[5] S. Marinsel, and G. Degardin, J. Eu.Cera. Soc. 21(10-11), 1919 (2001)

[6] P.G. Dennes Rev. Mod. Phys. 36, 225 (1964)

[7] J. Clarke, Proc. R. Soci. London, A 308, 442 (1969)

[8] D. Kumar, M.Sharon, P.R. Apte, S.P. Pai, R. Pinto, Appl. Phys. Lett. 61, 2105 (1992)

[9] V. Ambegaokar and A. Baratoff Phys. Rev. Litt 10, 486 (1963)

KEYWORD AND AUTHOR INDEX